AF361888

Air Pollution Meteorology

Air Pollution Meteorology

Editors

Isidro A. Pérez

M. Ángeles García

MDPI • Basel • Beijing • Wuhan • Barcelona • Belgrade • Manchester • Tokyo • Cluj • Tianjin

Editors
Isidro A. Pérez
Department of Applied Physics
University of Valladolid
Valladolid
Spain

M. Ángeles García
Department of Applied Physics
University of Valladolid
Valladolid
Spain

Editorial Office
MDPI
St. Alban-Anlage 66
4052 Basel, Switzerland

This is a reprint of articles from the Special Issue published online in the open access journal *International Journal of Environmental Research and Public Health* (ISSN 1660-4601) (available at: www.mdpi.com/journal/ijerph/special_issues/APM).

For citation purposes, cite each article independently as indicated on the article page online and as indicated below:

LastName, A.A.; LastName, B.B.; LastName, C.C. Article Title. *Journal Name* **Year**, *Volume Number*, Page Range.

ISBN 978-3-0365-2553-2 (Hbk)
ISBN 978-3-0365-2552-5 (PDF)

Contents

About the Editors

Isidro A. Pérez

His scientific career began with photochemical oxidant measurements and modelling. The following research line was the management of RASS sodar data, which led to different papers. This device analyses the low boundary layer beyond the heights investigated by any meteorological tower. After that, CO_2 concentrations were studied at a rural site and several papers that combined RASS sodar observations with concentrations measured present the results of this research. Moreover, parametric and nonparametric procedures were used to investigate the daily and annual cycles of greenhouse gases. The analysis of air parcel trajectories was later considered, and the contrast between Atlantic and continental trajectories was observed on the greenhouse gas concentration recorded. Finally, his research is oriented towards the trend determination by parametric and nonparametric procedures and the origin of greenhouse gas outliers and their influence on this trend.

M. Ángeles García

Her research has included photochemical oxidant measurements and modelling. Special attention was devoted to ozone analysing temporal variations, air quality and the influence of meteorological variables. Another research topic in which she has participated emphasizes the greenhouse gases included in some projects with the aim of studying the climate variability and the need to quantify the sources and sinks of CO_2 of natural ecosystems. Different procedures have been applied to investigate temporal cycles and trends of greenhouse gases, the influence of the air masses and the atmosphere evolution on the concentrations recorded. The research has also included the study of CO_2 fluxes in an agricultural soil.

Preface to "Air Pollution Meteorology"

A knowledge of human influence on pollutant concentrations in the atmosphere is both a current and a major issue, with the growth of these concentrations proving key to quantifying possible risks for both the population and materials. The air pollution system consists of three parts: the first is the sources that release these substances, the second is the atmosphere, which is a substratum where they evolve, while the receptors that experience the consequences of air pollution are the third part of this system. The atmosphere is beyond human control and is where chemical reactions as well as pollutant transport and removal processes, such as deposition, occur. These processes may be significantly conditioned by meteorological variables.

Air pollution covers a wide spatial range, and certain episodes present a local scope. For instance, ozone formation is favoured under high-pressure systems, with clear skies, a hot and dry atmosphere, strong solar radiation, and weak convection. Moreover, meteorological variables, such as relative humidity, may determine the type of air pollution, such as wet haze, which is linked to high relative humidity, and dry haze, when relative humidity is low.

However, pollution transport may extend the spatial range of this problem. Although transport may help to decrease air pollution at a local level due to the inverse relationship between concentrations and wind speed, certain substances, such as particulate matter, are sometimes transported from sources like deserts to sites where they could determine pollution episodes. They may even be accompanied by toxic substances, since some sites swept by air parcel motion are war zones. In some situations, microorganisms have even been transported, and avoiding outdoor activities for at least two days after these episodes is recommended. The air parcel path conditions this transport. For instance, photochemical ozone production is amplified by polluted and hot air coming from certain directions, whereas low ozone concentrations are linked to clean, cool, and dry air from a different site. Moreover, although transport is determined by orographic features, the wind may sometimes overcome these barriers.

Various procedures are used to quantify the influence of meteorology on air pollution, such as the direct relationship between meteorological variables and the concentrations of certain substances. However, a more elaborate treatment of observations is sometimes required, such as the analysis of synoptic meteorological patterns. Chemical-transport models are complex and help to supplement procedures aimed at investigating air pollution events under varied conditions. In particular, sensitivity analyses can be performed to investigate model response to changes in input variables. Moreover, meteorological conditions that favour high concentrations may be investigated.

This link between meteorology and air pollution is usually considered to be of secondary importance, yet merits additional study in order to enhance current knowledge. The result of this research will be useful not only from a theoretical point of view, i.e., in terms of gaining insights into atmospheric processes, but also with regard to implementing actions in the field of air pollution control. Consequently, exploring what impact meteorology has on air pollution processes is a promising research field, with varied study options and whose results have a direct social application. This book is aimed at improving current knowledge of this relationship. Its content includes a range of meteorological variables and synoptic patterns to describe atmospheric evolution. Air pollution is characterised by particulate matter, harmful substances and indoor air quality. Finally, the relationship between them is established by statistical procedures such as cluster analysis or multiple linear regression, while calculations such as air parcel trajectories are also considered. The result is a book that presents current research on the subject and which is geared towards increasing and sharing experience in this field with scientists and managers responsible for taking decisions to prevent or deal with air pollution situations.

Isidro A. Pérez, M. Ángeles García
Editors

 International Journal of *Environmental Research and Public Health*

Article

Pollution Sources and Carcinogenic Risk of PAHs in PM$_1$ Particle Fraction in an Urban Area

Ivana Jakovljević [1,*] , Zdravka Sever Štrukil [1], Ranka Godec [1], Ivan Bešlić [1], Silvije Davila [1], Mario Lovrić [2] and Gordana Pehnec [1]

[1]   Institute for Medical Research and Occupational Health, Ksaverska cesta 2, 10000 Zagreb, Croatia; zsever@imi.hr (Z.S.Š.); rgodec@imi.hr (R.G.); ibeslic@imi.hr (I.B.); sdavila@imi.hr (S.D.); gpehnec@imi.hr (G.P.)
[2]   Know-Center, Inffeldgasse 13, 8010 Graz, Austria; mlovric@know-center.at
*   Correspondence: ijakovljevic@imi.hr; Tel.: +385-1-4682589

Received: 30 September 2020; Accepted: 17 December 2020; Published: 21 December 2020

Abstract: Airborne particles are composed of inorganic species and organic compounds. PM$_1$ particles, with an aerodynamic diameter smaller than 1 μm, are considered to be important in the context of adverse health effects. Many compounds bound to particulate matter, such as polycyclic aromatic hydrocarbons (PAH), are suspected to be genotoxic, mutagenic, and carcinogenic. In this study, PAHs in the PM$_1$ particle fraction were measured for one year (1/1/2018–31/12/2018). The measuring station was located in the northern residential part of Zagreb, the Croatian capital, close to a street with modest traffic. Significant differences were found between PAH concentrations during cold (January–March, October–December) and warm (April–September) periods of the year. In general, the mass concentrations of PAHs characteristic for car exhausts (benzo(ghi)perylene (BghiP), indeno(1,2,3-cd)pyrene (IP), and benzo(b)fluoranthene (BbF)) were higher during the whole year than concentrations of fluoranthene (Flu) and pyrene (Pyr), which originated mostly from domestic heating and biomass burning. Combustion of diesel and gasoline from vehicles was found to be one of the main PAH sources. The incremental lifetime cancer risk (ILCR) was estimated for three age groups of populations and the results were much lower than the acceptable risk level (1×10^{-6}). However, more than ten times higher PAH concentrations in the cold part of the year, as well as associated health risk, emphasize the need for monitoring of PAHs in PM$_1$. These data represent a valuable tool in future plans and actions to control PAH sources and to improve the quality of life of urban populations.

Keywords: BaP; HPLC; carcinogenic; diagnostic ratio

1. Introduction

Particulate matter (PM) is assumed to be among the most hazardous of all ambient pollutants. Particle pollution contains "inhalable coarse particles" with diameters larger than 2.5 μm and smaller than 10 μm and "fine particles" with diameters of 2.5 μm or smaller. One of the most significant organic groups bound to PM in terms of health risk are polycyclic aromatic hydrocarbons (PAH). These compounds can exist in the atmosphere in the vapor phase (PAHs with low molecular weight), whereas heavier ones (PAHs with high molecular weight) are mostly adsorbed on the particle phase [1]. PAHs generally occur as complex mixtures, products of incomplete combustion processes, and originate from natural and anthropogenic sources [2]. High PAH levels in the ambient air of large metropolitan cities are usually linked with traffic, as well as diesel and gasoline automobiles [1,3]. Catalytic converters have shown a significant effect on reducing levels of the PAH concentrations in exhaust gases, but PAH emission levels continue to increase due to the contribution of other sources, such as traffic congestion [4]. PAHs are always emitted as a mixture, and the molecular concentration ratios are considered to be typical for a given emission source [5]. The toxicity, carcinogenicity, and mutagenicity of aromatic hydrocarbons have led to increased concerns for human populations. Long-term exposure to PAHs can cause toxic effects such as breathing problems, lung function abnormalities, decreased immune function, kidney and liver damage, skin irritation, and inflammation. The most significant health effect to be expected from inhalation of PAHs is increased risk of lung cancer primarily in occupations with heavy exposure to traffic-related air pollution, such as policeman [6] and newsagents [7]. Piccardo et al. [8] noticed that taxi drivers in Genoa were exposed to significantly higher daily BaP concentrations in comparison to workers from another occupational category. Fuel and biomass combustions, traffic emissions, and use of lubricant oils were identified as the main sources of PAH exposures at five Portuguese fire stations. During a normal shift in non-fire situations, levels of light PAHs were predominant and may promote some adverse health outcomes [9]. Bladder cancer is also linked to exposure to PAHs. Boada et al. [10] conducted a case study in the Canary Islands measuring PAH serum levels of 140 patients diagnosed for bladder cancer. Their results showed a difference in PAH contamination profile between patients and the control group, leading to the conclusion that specific PAH mixtures play a role in bladder cancer. People are persistently exposed to PAHs. Benzo(a)pyrene (BaP) is the most investigated PAH, and most information on the toxicity and manifestation of PAHs is related to this compound, which is why it was used as an indicator of carcinogenic hazard in polluted environments [11–13].

Although particle composition has been determined by many authors, there is still not much research related to the study of PAHs in airborne fine particles. Fine particles, with an aerodynamic diameter of less than 2.5 μm ($PM_{2.5}$), and especially particles with aerodynamic diameter of less than 1 μm (PM_1), may play an important role in affecting human health. Squizzato et al. [14] stated the following reasons: they penetrate more effectively into the deep lung; they can penetrate more readily into indoor environments; they can remain suspended for longer periods of time in the atmosphere than coarse particles; they may be transported over long distances; they tend to carry higher concentrations of more toxic compounds, including acids, heavy metals, and organic compounds; and they have a larger surface area per unit mass compared to larger particles and can thus absorb larger amounts of semi-volatile compounds. Measurements of PM_1 and its content have so far not been included in routine measurement programs, although they would provide more information about potential PM sources and could be used to improve PM control strategies and health protection. A recent study by Yang et al. [15] found that both PM_1 and $PM_{2.5}$ levels were associated with poorer lung function in children, with stronger associations for PM_1 compared to $PM_{2.5}$, pointing to the importance of regulating finer PM fractions. Furthermore, a lot of previous health risk estimations were carried out taking into account larger fractions of particulate matter (PM_{10} or even total suspended particles), which do not enter deeply into the respiratory system [16–18].

Studies regarding PAH in PM_1 were carried out only at a few urban locations in Europe [19–25]. It was found that the levels varied significantly between heating and non-heating season [22,23,26]. The highest concentrations were observed during winter in areas where coal and wood burning were used for heating [22,24]. Traffic-loaded sites showed a large contribution of PAHs with larger molecular weight [21,22,27,28].

In this study, the mass concentrations of individual PAHs in the PM_1 particle fraction were determined in an urban area together with some gaseous pollutants. The relationship between measured species and meteorological parameters (temperature, relative humidity, atmospheric pressure, wind direction, and velocity) was determined. Potential pollution sources were assessed using PAH diagnostic ratios, Spearman's regression, and principal component analysis (PCA). In a previous study at the same location [19], factor analysis (FA) was used to identify pollution sources. The aim of both PCA and FA is to determine the main relationships between the observed variables and to reduce the large number of observed variables to a smaller number of factors. Mathematically, in the PCA, the entire variance in the observed variables is analyzed, while in the FA only the mutual variance is analyzed. Attempts are made to identify and eliminate variance due to error and variance specific to each individual variable. The assumption of FA is that factors "create" variable values, while in PCA it is assumed that variables "create" components. In PCA, there is no basic theory that would define how variables are grouped into factors, the variables are simply empirically related within components. As atmospheric PAH levels are affected by different weather conditions and numerous atmospheric reactions, and it is not possible to determine their interdependencies for each day, we estimated that the PCA method is more applicable for this analysis. In addition, the adverse health effect of PAHs on a human organism was assessed using toxic equivalent concentrations. The incremental lifetime cancer risk (ILCR) was estimated for three age group of populations, which presents the first such study carried out for PAHs in PM_1 in this part of Europe.

2. Materials and Methods

2.1. Location and Sampling

Concentrations of 11 PAHs in PM_1 particle fraction were measured continuously from January to December 2018, together with gaseous pollutants (CO, NO_2, SO_2, and O_3). The measuring station (45°50′6.83″ N, 15°58′42.12″ E, 168 m a.s.l.) was situated in the northern residential area of Zagreb, the Croatian capital (~800.000 inhabitants). It is a low-rise building area with small inhabitant density and mild traffic density. Residential (domestic) heating relies mostly on gas, but some households still use oil or wood for heating and cooking.

Samples of PM_1 particle fraction were collected on quartz filters with low-volume sequential automatic sampler (Sven Leckel) from about 55 m^3 of air per day. Filters were collected 24-h a day, and the total number of samples was 363. The sampler inlet was located approximately 1.5 m above ground and 15 m away from the road. The PM_1 samples were kept frozen in aluminum foil at −18 °C until PAH analysis to avoid PAH losses and sample degradation. The filters were extracted no later than two months after sampling.

2.2. Measurements of Gaseous Pollutants and Meteorological Data

Gaseous pollutants were measured using automatic devices. SO_2 was determined using a HORIBA APSA 370 device according to the norm (EN 1421:2014). NO_2 was measured by a HORIBA APNA 370 device (according the norm EN 14211:2012), O_3 was measured using a HORIBA APOA 370 device (according the norm EN 14625:2012), and CO was measured using a HORIBA APMA 370 device (according the norm (EN 14626:2012). Measurements of SO_2, CO, NO_2, and O_3 are part of the local air quality monitoring network funded by the City of Zagreb, City Office for Economy, Energetics and Environment Protection.

Meteorological data (temperature, relative humidity, pressure, amount of rainfall, wind direction, and velocity) were obtained from the nearest meteorological station (Maksimir) of the Croatan Hydrological and Meteorological Service (www.meteo.hr).

2.3. Analysis of PAHs

Extraction of PAHs from filters was performed with a solvent mixture (toluene:cyclohexane 7:3) in an ultrasonic bath. Further preparation included centrifugation (10 min, 3000 rpm) and evaporation to dryness. After that, samples were redissolved in acetonitrile. Concentrations of PAHs were determined by Agilent Infinity 1260 high-performance liquid chromatography (HPLC) with a fluorescence detector. Zorbax Eclipse PAH column (100 × 4.6 mm) was used for separations of PAHs. The mobile phase was acetonitrile and water (60:40), with a flow rate of 1 mL min^{-1} [29,30]. Calibration curves were prepared with a commercial PAH standard (Supelco EPA 610 PAHs Mix). The calibration range was from 0.005 ng μL^{-1} to 0.08 ng μL^{-1} for pyrene, benzo(a)anthracene, chrysene, benzo(k)fluoranthene, benzo(a)pyrene, and indeno(1,2,3,cd)pyrene, but for fluoranthene, benzo(j)fluoranthene, benzo(b)fluoranthene, dibenzo(ah)anthracene, and benzo(ghi)perilene the calibration range was from 0.01 ng µL^{-1} to 0.16 ng µL^{-1}. Samples were analyzed for the following PAHs: fluoranthene (Flu), pyrene (Pyr), benzo(a)anthracene (BaA), chrysene (Chry), benzo(j)fluoranthene (BjF), benzo(b)fluoranthene (BbF), benzo(k)fluoranthene (BkF), benzo(a)pyrene (BaP), dibenzo(ah)anthracene (DahA), benzo(ghi)perylene (BghiP), and indeno(1,2,3-cd)pyrene (IP). The limit of detection (LoD) was from 0.001 ng m^{-3} for BaA to 0.03 ng m^{-3} for BjF. The quantification limit (QL) varied from 0.002 ng m^{-3} for BaA to 0.1 ng m^3 for BjF. The accuracy of the method was determined by analyzing the standard reference material (SRM 1649a, urban dust) provided by the National Institute of Standards and Technology (NIST) and ranged from 88% for Flu to 109% for BkF. Samples of SRM 1649a were processed the same way as real samples. The details of detection and quantification limits, as well as accuracy are shown in Table S1 of the Supplementary Data section.

2.4. Statistical Analysis

The statistical results were processed by Microsoft Excel and the Statistica 13 (Tibco Software Inc.) program. Statistical significance was set at 5% ($p < 0.05$). The Shapiro-Wilk test was used to test the normality of variables. Seasonal differences between warm and cold period for each PAH concentrations, heavy and light PAHs, and PM$_1$ particles were tested by Mann–Whitney U test.

The concentration ratios between selected PAHs were investigated to find out more about the nature of pollution sources. Diagnostic ratios are common tools for the identification of pollution sources [30–33]. However, some papers have presented their restrictions. Katsoyiannis et al. [34] demonstrated that diagnostic ratios cannot be effective as markers of sources, because they are influenced by weather condition and atmospheric reaction of PAHs such as photodegradation, and reaction with ozone and other atmospheric pollutants. To minimize confounding factors such as dissimilarities in volatility, water solubility, and adsorption, diagnostic ratio calculations usually are restricted to PAHs of similar molecular mass [35]. Due to the restrictions of the diagnostic ratio, the multivariate principal component analysis (PCA) was performed on data and calculated anomalies for the studied samples. Principal components (PCs) with eigenvalues greater than 1 for both periods were calculated, and their contributions in the total variance were determined. We graphically displayed the PCs loading as well as the orientation of the variables and samples with respect to these principal components.

2.5. Carcinogenic Activity and Population Exposure

Benzo(a)pyrene was used as an indicator of health impact of the PAH mixture to human health. Many studies have shown that BaP was present at more than 50% in total carcinogenic activity and that made it a good indicator of carcinogenic activity of the PAH mixture. The carcinogenic activity of individual PAHs was estimated on the basis of toxic equivalency factors (TEFs) from the literature. Different TEF schemes are developed by different authors, based on experiments in animals [36–39]. Nisbet and LaGoy [39] completed a new list of TEFs, which appears to better reflect the current state of knowledge on the relative potency of individual PAHs. In this study we used Nisbet and LaGoy's [39] toxic equivalent factors.

For calculating the risk of the PAH mixture in ambient air, the carcinogenic potencies of individual PAHs are expressed relative to the potency of BaP. BaP equivalents (BaP$_{eq}$) were calculated by multiplying the mass concentration of an individual PAH with its respective toxic equivalency factor. Total carcinogenic potency (TCP) was calculated by summing up the BaP$_{eq}$ of each measured PAH. The equation used to calculate the TCP is presented below (1):

$$TCP = \Sigma BaP_{eq} = \Sigma \gamma i \cdot TEFi \tag{1}$$

TCP—total carcinogenic potency/ng m^{-3}
BaP$_{eq}$—BaP equivalents concentration
TEFi—toxic equivalency factor of a particular PAH
γi—mass concentration of individual PAH/ng m$^{-3.}$

To determine the daily population exposures, total carcinogenic potency was used to calculate the daily dose according to Equation (3). In this study we tried to estimate the most probable scenario for three age groups: infant (0–1 year), children (5–19 year), and adult (20–70 year). If we assumed that people spent an average of ten hours at their job/school and eight to ten hours at home (including sleeping), we can assume that they were elsewhere for the rest of their day, due to the fact that people spend approximately 25% of their time outdoors, that is 6 h per day.

The incremental lifetime cancer risk (ILCR) posed by exposure to PM_1-bounded PAHs was computed following Equation (2) [20].

$$ILCR = ((SF_{inh} \times IEL \times EF \times ED)/(BW \times AT \times cf)) \times y \tag{2}$$

SF_{inh}—inhalation cancer slope factor of BaP/kg day mg^{-1}

IEL—BaP_{eq} daily dose/ng day^{-1}

EF—exposure frequency/day $year^{-1}$

ED—daily exposure level/μg g^{-1}

BW—body weight/kg

AT—average time/day

Cf—conversion factor (10^{-6})

y—age

In this study, SFinh, EF, and ED was used as derived by Chen and Liao [17]. Parameters were different for infants, children, and for adults, with average body weights of 6.79 kg, 36.24 kg, and 59.78 kg, respectively. Parameters that are used in calculation of ILCR are shown in Table S3 of the Supplementary Data section. Results of ILCR shows incremental cancer risk per year and for lifetime cancer risk this result should be multiplied by the age of the person.

In this study, a SFinh of 3.14 kg day mg^{-1} was used, and the proper description of how this value was derived was explained by Chen and Liao [17]. The BaP_{eq} daily dose was calculated by multiplying the TCP concentrations (ng m^{-3}), inhalation rate (IR, m^3 day^{-1}), and daily exposure time span (t, h). The equation used to calculate the IEL is presented below (3):

$$IEL = TCP \times t \times IR \tag{3}$$

3. Results and Discussion

3.1. PAH Concentrations

The monthly average mass concentrations of the PM_1 particle fraction during the whole calendar year are shown in Figure 1. This result shows seasonal differences ($p < 0.001$) of PM_1 concentrations with high values during the cold period of the year (January–March; October–December) and lower values during the warm period (April–September). The average 24-h concentrations of PM_1 varied from 0.7 to 55.3 μg m^{-3} with an annual average of 13.6 μg m^{-3}. Monthly average concentrations were also calculated for the PAH sum and for individual PAHs, and the results are presented in Figures 2 and 3. Sums of PAHs concentrations (ΣPAHs) ranged from 0.437 ng m^{-3} during June to 21.497 ng m^{-3} during December, and the annual mean mass concentration was 6.354 ng m^{-3}. The highest mass concentration during the cold period was measured for BbF, while in the warm period, the highest mass concentrations were determined for BghiP. High BghiP mass concentrations in the warm period in comparison to other hydrocarbons were probably due to BghiP stability at high temperatures. The lowest mass concentrations for both periods (cold and warm) were determined for DahA. The range of the monthly average concentrations for DahA was from 0.010 ng m^{-3} (May) to 0.386 ng m^{-3} (December). The average monthly mass concentrations of BaP ranged from 0.038 ng m^{-3} (June) to 2.826 ng m^{-3} (December), while the annual mass concentrations were 0.765 ng m^{-3}. The target value in the European Union set by Directive 2004/107/EC for BaP content in the PM_{10} fraction is only 1 ng m^{-3} averaged over a calendar year. The reason for higher concentration exclusively during the cold part of the year can be explained because of the long temperature inversion periods that occur in Zagreb. Such weather conditions are characterized by high pressure and impaired air mixing, which favors the pollutants accumulation. These stable atmospheric conditions combine with increased emissions from heating are the basic origins of elevated concentrations of PAHs during the cold season. In the study area, gas, oil or, wood are mostly used for domestic heating, while coal has not been in use for more than thirty years.

The relevant literature comprises a very limited number of papers related to PAHs in the PM$_1$ fractions [19–24,27,28,40,41]. Table S2 of the Supplementary Data section shows the summary results of PAH mass concentrations obtained in this study in comparison with similar studies. Rogula-Kozłowska et al. [40], Kozielska et al. [22], and Majewski et al. [20] reported much higher mass concentrations of PAHs in the PM$_1$ particle fraction during both the heating and non-heating seasons at different locations in Poland compared with Zagreb. Much higher winter concentrations were also found in Ostrava-Radvanice (industrial site), in Kladno-Švermov and Brno (urban sites) in Czech Republic, while levels of PM$_1$ and ΣPAHs were similar in Košetice and Čelakovice. In a coastal area in Poland (Baltic Sea), BaP levels were much higher than in this study, while in a coastal area in Greece PM$_1$ the levels were similar as in Zagreb but particle-bound PAHs were significantly lower. Similar PAH concentrations were also found in Guadalajara Metropolitan Area in Mexico while PAH concentrations determined in the Metropolitan Area of Porto Alegre, Brazil, were much lower during winter but higher during summer [28].

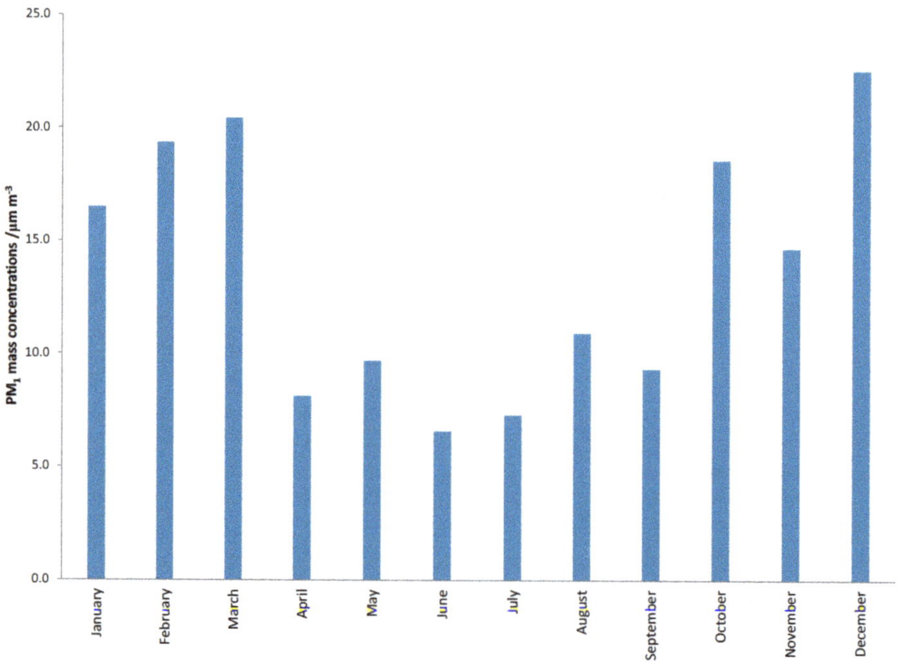

Figure 1. Monthly mass concentrations of the PM$_1$ particle fraction during one calendar year.

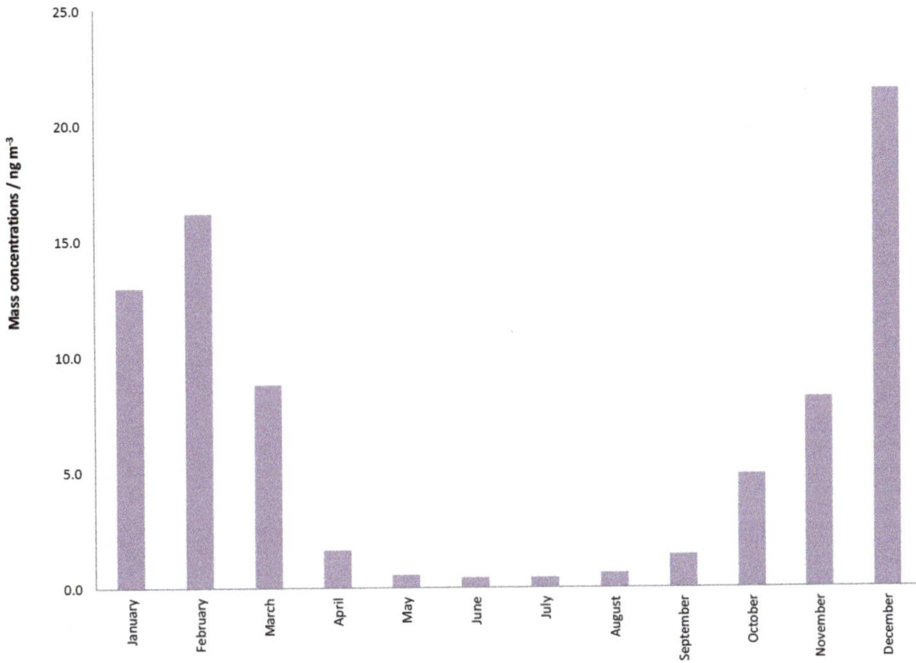

Figure 2. Monthly mass concentrations of ΣPAHs (polycyclic aromatic hydrocarbons) measured during one calendar year.

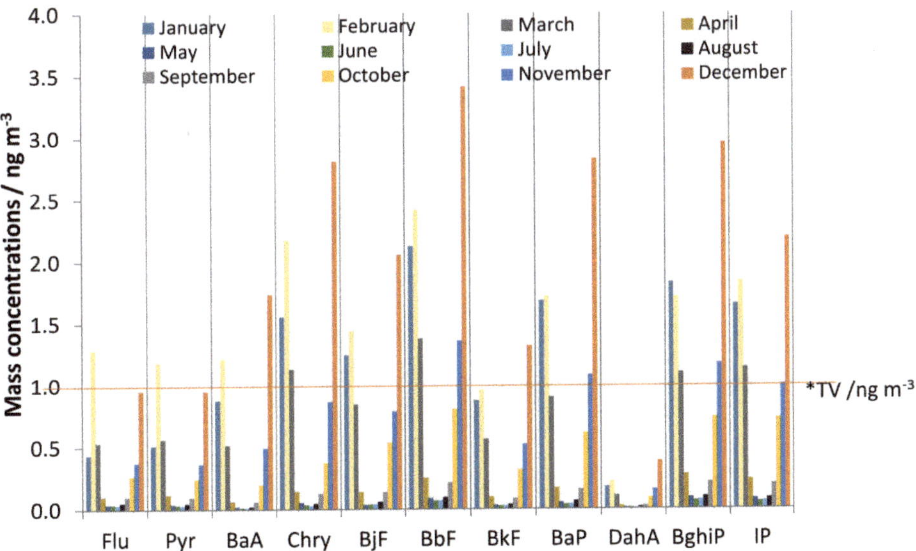

Figure 3. Monthly mass concentrations of PAHs measured during one calendar year. * The target value (TV) in the European Union set by Directive 2004/107/EC for BaP content in PM_{10} fraction is 1 ng m^{-3} averaged over a calendar year.

As significant differences were found between cold and warm months, we divided the results into two groups, separately for the cold (January–March; October–December) and the warm (April–September) period of the year. Summary statistical parameters for all of the measured PAHs for these two periods are shown in Table 1. Concentrations of PAH characteristic for domestic heating or biomass burning (Flu, Pyr) were lower in the cold period than those of PAHs characteristic for car exhausts (BghiP, BbF, and IP). This indicated that the PAHs in the PM_1 particle fraction could originate predominantly from car exhausts [42–44].

Table 1. Average, minimum, and maximum values and standard deviations of the PAH 24-h mass concentrations during the cold and warm period (ng m^{-3}).

	Warm				Cold			
	C_{min}	C_{max}	C	SD	C_{min}	C_{max}	C	SD
$N = 182$					$N = 181$			
Flu	0.009	0.362	0.063 *	0.051	0.041	3.718	0.631 *	0.622
Pyr	0.007	0.358	0.062 *	0.056	0.040	2.896	0.627 *	0.560
BaA	0.008	0.302	0.033 *	0.038	0.051	5.535	0.828 *	0.829
Chry	0.016	0.619	0.074 *	0.078	0.106	8.338	1.464 *	1.320
BjF	0.010	0.579	0.076 *	0.082	0.123	5.209	1.141 *	0.845
BbF	0.027	0.852	0.131 *	0.125	0.228	8.600	1.894 *	1.389
BkF	0.010	0.340	0.052 *	0.051	0.089	3.185	0.751 *	0.539
BaP	0.011	0.665	0.088 *	0.098	0.150	7.483	1.454 *	1.216
DahA	0.004	0.087	0.017 *	0.013	0.021	1.003	0.189 *	0.155
BghiP	0.022	0.792	0.137 *	0.134	0.235	7.027	1.556 *	1.116
IP	0.019	0.748	0.120 *	0.120	0.231	5.250	1.413 *	0.897
ΣPAH	0.177	5.705	0.852 *	0.828	1.384	57.146	11.815 *	9.082

C_{min}—minimum value; C_{max}—maximum value; C—arithmetic mean; SD—standard deviation; and N—number of samples. * Seasonal differences were tested by Man–Whitney U test ($p < 0.001$).

According to the number of rings and molecular weight, PAHs can also be classified into two groups: heavy and light. Heavy PAH concentrations were calculated as the sum of PAHs with five or more aromatic rings, and light PAHs represent the sum of PAHs with four aromatic rings. Heavy PAHs are usually characteristic for car exhausts, while light PAHs originate mostly from domestic heating or biomass burning [43,44] (Figure 4). Figure 4 shows the monthly mass concentration for these two groups. During both measurement periods, the contributions of heavy PAHs were much higher (>69%) than those of light PAHs (Table 2). Concentrations of PM_1-bounded PAHs during warm periods can easily evaporate from particles to gas phase, so their concentrations in the gas phase increased, but in the particle phase their concentrations decreased. The reason behind this could be that for these PAHs the dominant sources were biomass burning and during the warm period the effect of these sources were minimal. These results also indicated that traffic (diesel or gasoline) could be the main pollution source of PAHs in PM_1 in this area. To confirm the assumption, we performed diagnostic ratio and principal component analysis to identify possible pollution sources.

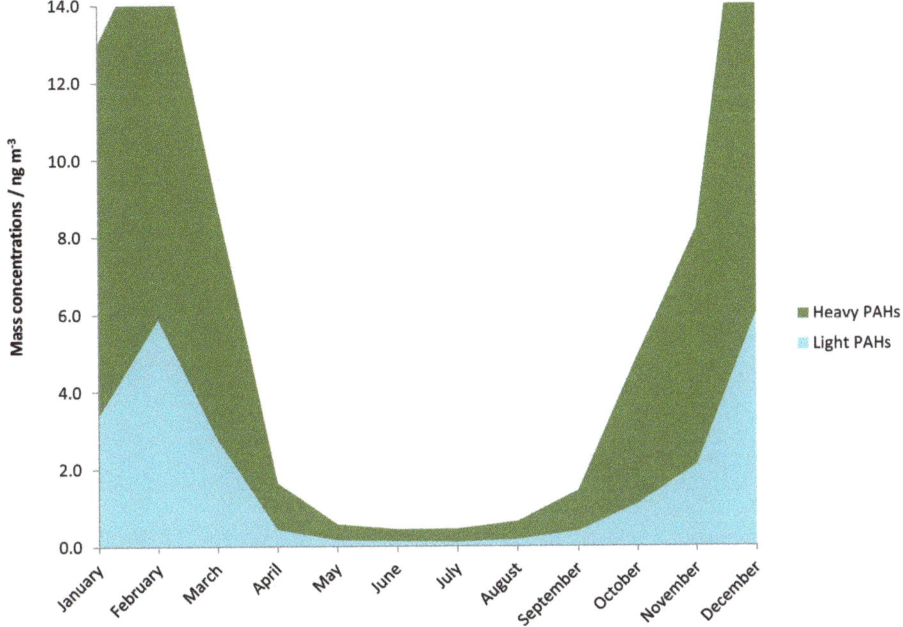

Figure 4. Mass concentrations of light and heavy PAHs measured during one calendar year.

Table 2. Contribution of light and heavy PAHs in the sum of measured PAHs during cold period and warm period.

Parameter	Light PAH		Heavy PAH	
	Cold	Warm	Cold	Warm
Average/ng m^{-3}	3.516 *	0.232	8.308 *	0.620
Contribution/%	29.4	27.2	69.5	72.8

* Seasonal differences between warm and cold period (test by Mann–Witney U test; $p < 0.001$).

3.2. Diagnostic Ratios

Specific ratios of individual PAHs are characteristic for some combustion processes and many authors used diagnostic ratios of PAHs to identify potential pollution sources [10–13]. In this study, the following ratios were selected: IP/(IP + BghiP), BaP/BghiP, Flu/(Flu + Pyr), and BaP/(BaP + Chry). The values determined in this study were compared with the same diagnostic ratios computed from characteristic emission sources based on the relevant literature. IP/(IP + BghiP) ratio values between 0.35 and 0.7 are characteristic for diesel, while some authors reported a value of 0.56 that indicates coal combustion [28,45]. BaP/BghiP values between 0.3 and 0.4 are characteristic for traffic, 0.46–0.81 for diesel combustion, and 0.9–6.6 for coal combustion [28,31,46]. A Flu/(Flu + Pyr) ratio between 0.2 and 0.5 indicates diesel, a ratio between 0.4 and 0.5 liquid fossil fuel, while a ratio >0.5 suggests wood combustion [32,33]. Finally, a BaP/(BaP + Chry) ratio of <0.5 indicates diesel but >0.5 gasoline [4,28]. Average PAH diagnostic ratios for the cold and warm period are presented in Table 3.

Table 3. Comparison of PAH average diagnostic ratios in PM_1 during cold and warm period and the main emission sources.

Diagnostic Ratio	Literature Data			This Study		
	Ratio Value	**Source**	**Reference**	**Average**	**Percentile**	**Potential Source**
Flu/(Flu + Pyr)	0.2–0.5 0.4–0.5 >0.5	diesel liquid fossil fuel wood combustion	[4] [33] [32]	Cold 0.5 Warm 0.5	96 85	diesel, liquid fossil fuels
BaP/BghiP	0.3–0.4 0.46–0.81 0.9–6.6	traffic diesel coal or wood combustion	[28] [46] [31]	Cold 0.9 Warm 0.6	60 67	diesel/wood diesel
BaP/(BaP + Chry)	<0.5 >0.5	diesel gasoline	[4] [28]	Cold 0.5 Warm 0.5	70 60	diesel/gasoline
IP/(IP + BghiP)	0.35–0.70 0.56	diesel coal	[28]	Cold 0.5 Warm 0.5	97 99	diesel

In this paper, the average IP/(IP + BghiP) ratios were 0.5 during both the cold and warm periods, suggesting that the produced PAHs stemmed from the emission of diesel vehicles. Those average values (0.5) for cold and warm period also corresponded to the 97th and 99th percentile value, respectively, i.e., during the cold period 97% of days had a ratio value lower than 0.5, while during the warm period 99% days were with a ratio of <0.5. The average BaP/(BaP + Chry) and Flu/(Flu + Pyr) ratios were 0.5 during both periods. The average ratio BaP/(BaP + Chry) in the same time represents the 60- and 67-percentile for cold and warm periods, respectively. The other ~40% of results were higher than average value, and the individual value did not exceed 0.7, indicating both diesel and gasoline combustion as potential sources. The average ratio value Flu/(Flu + Pyr) was marginal between biomass burning, emissions from gasoline or diesel vehicles, and combustion of other liquid fossil fuels, but during the warm period there were more days (15%) with a ratio value characteristic for wood burning (during the cold period 96% of results was <0.5). Because of that, it could be concluded that during the warm period mixed sources are probably present (wood combustion and emissions from gasoline or diesel vehicles). The BaP/BghiP ratio was 0.9 during the cold and 0.6 during the warm period. In the warm period, the value was characteristic for emission from diesel (Table 3), but in the cold period it was characteristic for mixed sources (coal, wood combustion, and emission from diesel). The BaP/BghiP ratio included 70% and 60% results during the cold and warm period, respectively. The other 30% during the cold period were higher than 0.9, which indicated that coal or wood combustion were present as pollution sources. During the warm period, 40% of results were higher than 0.6 but did not exceed 0.8, which is the highest value for diesel combustion. Similar results were reported by Agudelo-Castañeda and Teixeira [28] and Hanedar et al. [46]. They also found that PAHs in PM_1 originated predominately from diesel or gasoline emission. The contributions of the individual PAHs to the sum of total PAH mass concentration were calculated as well and are presented in Figure 5.

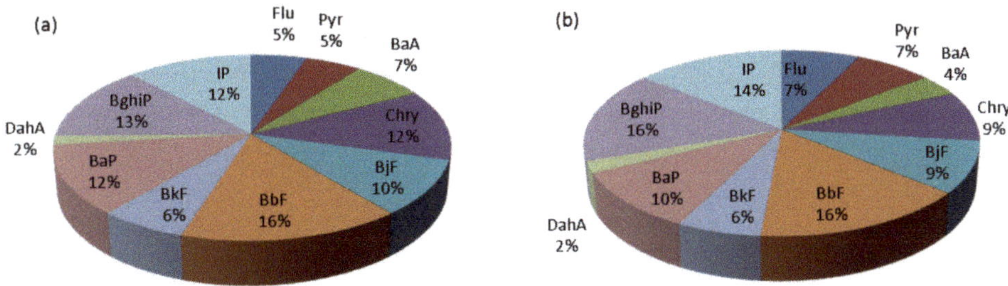

Figure 5. The contributions of the individual PAHs to the sum of total PAH mass concentration during: (**a**) cold and (**b**) warm period.

From Figure 5 it is evident that the PAH mixture was characterized by a high contribution of 6-ring (BghiP, IP) and 5-ring PAHs (BaP, BjF, BbF, BkF) characteristic for vehicle exhaust emission. Contributions of Flu and Pyr (markers for biomass burning) in the cold period were similar to warm period and it was 5% in cold and 7% in warm period. Previous investigations at the same measuring site but in the PM_{10} particle fractions showed similar results during heating season [26]. These results suggest that four-ring PAHs concentrations have similar dominant sources during the warm and cold period (probably emission from gas and vehicle exhaust and domestic heating). A higher result than those measured in this study was found in Sarajevo [47] and in Poland [40]. Results in Sarajevo and Poland showed a higher contribution of four–ring PAHs (Flu, Pyr, and BaA), which were emitted from coal and wood combustion. The principal compound was BbF (16%) followed by BghiP (13–16%), IP (12–14%), and Chry (9–12%) in both periods. All of them were of pyrolytic origin, which suggests that PAHs reacted at similar extents and the dominant sources were similar.

However, Katsoyiannis et al. [34] presented that the diagnostic ratio cannot be effective as markers of pollution sources, because they can be influenced by atmospheric reaction of PAHs and weather conditions. Because of the fact that the diagnostic ratio can lead to mistakes in estimating the possible pollution sources, in this paper, PCA analysis was used in order to determine the most probable main pollution sources.

3.3. Principal Component Analysis

Furthermore, for investigating the similarities and differences between samples, PCA was applied, both for the cold and warm period. In the cold and warm period, the eigenvalues of the first two PCs were larger than 1, which indicated their significance. In the cold period, the first two principal components represented 98.08 cum.%, with the first and the second PCs contributing to the total variance of the data set with 89.09 cum.% and 8.99 cum.%, respectively. Results of the PCA during the cold period are presented on the PC1-PC2 loading (Figure 6a) and score plots (Figure 6b), illustrating the orientation of the variables and samples with respect to these principal components.

All variables had a negative effect on PC1, while all PAHs had positive values on PC2 except for Flu, Pyr, and Chry. Chry showed the lowest negative values. The highest negative effect on PC2 was for Pyr and Flu. In the warm period the first two principal components represent 97.74 cum.% whereas the first and the second PCs contributed with 95.20 cum.% and 2.54 cum.%, respectively, to the total variance of the data set. Results of the PCA during the warm period are presented on the PC1-PC2 loading (Figure 7a) and score plots (Figure 7b), illustrating the orientation of the variables and samples with respect to these principal components.

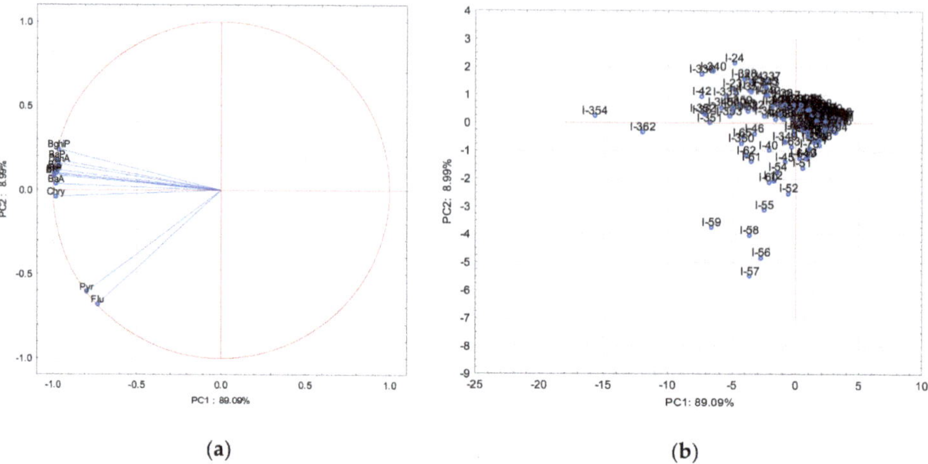

Figure 6. PC1-PC2 loading (**a**) and score plot (**b**) during cold period. I-1 to I-365 are the numbers of days.

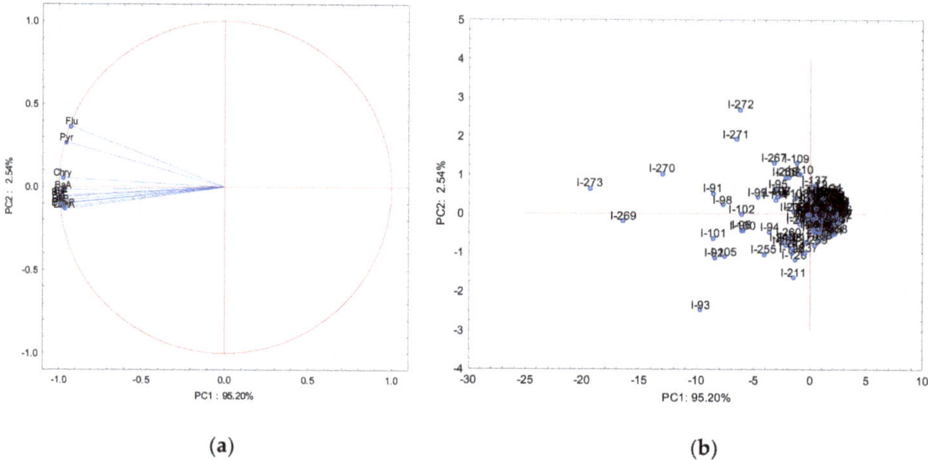

Figure 7. PC1-PC2 loading (**a**) and score plot (**b**) during warm period. I-1 to I-365 are the numbers of days.

All of the variables also had a negative effect on PC1 as in the cold period, while Flu, Pyr, and Chry exhibited positive values on PC2, but the rest of PAHs showed negative values. Based on these parameters, two different factors were extracted during both periods. In the cold period, Flu and Pyr were grouped separately because their common origin was different from the rest of the PAHs. These are PAHs characteristic for biomass burning, and their sources probably could have been domestic heating [4,28,48,49]. In the warm period, these two PAHs (Flu and Pyr) were extracted separately probably because they evaporated easily from particles, which makes them more sensitive to meteorological conditions than the other PAHs [50]. However, the results obtained by PCA analysis confirmed those obtained by the diagnostic ratio, but with limitations in the warm period when processes such as meteorological condition and evaporation come into play. PCA analysis secluded some samples during the warm and cold period. From the PC1-PC2 loading, it is evident that six samples during the warm period (Figure 7b) and seven samples during the cold period (Figure 6b) were secluded from other samples. Results of contribution of the individual PAHs to the sum of total PAH for only those selected days showed some differences from the contribution made for all measured warm and cold periods. During warm period samples, I-271 and I-272 had the highest contribution of Flu and Pyr, but samples I-93, I-269, I-270, and I-273 had the highest contribution of BaP except for the I-273 sample, which also had the highest contribution of Chry. During the cold period, samples I-55-59 had the highest contribution of Flu and Pyr, and samples I-354 and I-362 had the highest contributions of BaP, Chry, and BaA. At the end of February and beginning of March (samples I-58 to I-59), the air temperature was extremely lower (from -2 to -10 °C) and probably emission from domestic heating was higher, which caused a higher contribution of Flu and Pyr to the total PAH concentration. For days at the end of December (I-354 and I-362), the air temperature was not as extremely low as in February but nevertheless below zero, and these days are also celebratory days, which can cause high concentrations of PAHs. For the warm period, at the end of September (samples I-269 to I-273) the air temperature was approximately 10 °C, which is much lower than the average value for the warm period (20 °C). Results of PCA analysis were in very good agreement with diagnostic ratios, and for those days, the diagnostic ratios showed mixed sources (diesel/gasoline and wood combustion).

PCA analysis extracted two factors during the cold and warm periods: the first factor separated Flu and Pyr, and the second factor the rest of the PAHs. Because of that, wind roses were shown for two groups, the sum of Flu and Pyr concentrations (ΣFlu+Pyr) and the sum of the remaining nine PAHs (ΣRest PAHs). The dependence of PAH concentrations on wind direction is shown on Figure S1 of the Supplementary Data section together with wind frequencies and wind velocities.

Winds coming from ENE were the most frequent during the cold and warm periods, followed by winds from NNE and SE-ESE sector, while the winds from other directions were relatively rare. On the other hand, winds of high velocities came from north-western and south-western directions, as well as from NE-ENE and SE sections. In the cold months, the highest concentrations of ΣFlu+Pyr primarily came from the west (residential part with domestic heating and street with modest density) and then from the south (center of the town, with dense traffic, while in the warm months the highest concentrations came from the east (industrial part of the town). The difference in concentrations in the cold and warm part of the year is greater than 80%. For the ΣRest PAH wind roses look similar as for ΣFlu+Py, but in the cold months of the year the dominant direction was SSE and then W, while in the warm months the dominant source of pollution was the east (Figure S2). In the cold periods of the year, pollution concentrations from the east were completely absent, which indicated that in the cold months the dominant source of pollution came mostly from the west for ΣFlu+Pyr while for ΣRest PAH it can be concluded that the dominant source of pollution was located both in the west and in the SSE direction. Taking into account the general absence of winds from west direction it can be concluded that the dominant sources of PAHs are probably of local origin. In the warm months of the year, there was no direction from which there was no pollution.

3.4. Carcinogenic Activity of the PAH Mixture and Health Impact

The BaP equivalent concentration and total carcinogenic potency were calculated according to Equation (1) and are shown in Table 4. In the cold period of the year, the TCP was more than ten times higher than in the warm period. The average TCP in the warm period was 0.141 ng m^{-3} and in the cold 2.139 ng m^{-3}. These results were similar to the TCPs obtained previously in Zagreb [26]. In a study by Pehnec and Jakovljević [26], TCP values were shown for 30 samples collected per season (spring, summer, autumn, and winter) and TCP values varied from 0.063 ng m^{-3} during summer to 4.503 ng m^{-3} during winter. TCP values similar or lower than the ones in this study were reported by other authors who used the same TEF scheme [20,28], while higher TCP values were reported for some locations in Poland [22,51] and the Czech Republic [24].

Table 4. Equivalent BaP_{eq} (ng m^{-3}) concentrations for individual PAHs measured during the warm and cold period.

PAH	TEF [a]	BaP$_{eq}$ Warm	Cold
Flu	0.001	0.0001	0.001
Pyr	0.001	0.0001	0.001
BaA	0.1	0.003	0.082
Chry	0.01	0.001	0.014
BjF	-	-	-
BbF	0.1	0.013	0.187
BkF	0.1	0.005	0.074
BaP	1	0.088	1.438
DahA	1	0.017	0.187
BghiP	0.01	0.001	0.015
IP	0.1	0.012	0.140
TCP [b]		0.141	2.139
% BaP		62.2	67.2

[a] according to Nisbet and LaGoy (1992) [b] total carcinogenic potency.

Contributions of BaP to the carcinogenic potency in this study exceeded 60%. These values were similar for both measuring periods and were on average 62% and 67% for the warm and cold season, respectively. This indicates that benzo(a)pyrene had an important role in the carcinogenic activity of the PAH mixture. The same contribution was reported by Jakovljevic et al. [29] for the same location. Many authors reported similar BaP contributions but in other (PM$_{10}$ or PM$_{2.5}$) particle fractions [46,52].

To determine the daily population exposures, total carcinogenic potency was used to calculate the daily dose according to Equation (3). In this study, we tried to estimate the most probable scenario for three age groups: infant (0–1 year), children (5–19 year), and adult (20–70 year). If we assumed that people spent an average ten hours on their job/school and eight to ten hours at home (including sleeping), we can assume that they were elsewhere for the rest of their day, due to the fact that people spend approximately 25% of their time outdoors.

The incremental lifetime cancer risk (ILCR) posed by exposure to PM-bounded PAHs was computed following Equation (2) [20].

We used Equation (3) for calculations of IEL for the warm and cold period, for three age group and the results are shown in Table 5.

Table 5. The BaPeq daily dose (IEL) and the incremental lifetime cancer risk (ILCR) for three age groups.

Parameter	Infant Warm	Cold	Children Warm	Cold	Adult Warm	Cold
Age	0–1		5–19		20–70	
IEL/ng day^{-1}	5.04	76.5	21.0	319.2	27.7	420.2
ILCR	2.21×10^{-11}	3.35×10^{-10}	3.01×10^{-10}	4.56×10^{-9}	7.13×10^{-10}	1.08×10^{-8}

As the variables in the ILCR calculation belong to certain distributions taking the average value into account can be misleading for yielding the risk estimation. Therefore, we employed a probabilistic Monte Carlo (MC) simulation to feed the distributions to the ILCR model for infants, children, and adults in cold and warm seasons. All of the sampled variables were sampled randomly and independently. The information on variable distributions was taken from Chen and Lia [17] and Liu et al. [18]. We sampled the TCP from our own data after log-transforming them, separately for warm and cold seasons. The MC was iterated 10,000 times per sample variable and finally for the IEL and ILCR models. The parameter values used for MC are show in Table S3 of Supplementary Data. The average ILCR value and the MC simulation results for ILCR were than compared.

IEL was lower during the warm than cold period; 5.04 ng day^{-1}, 21.0 ng day^{-1}, and 27.7 ng day^{-1} in the warm period for infants, children, and adults, respectively. In the cold period, the BaPeq daily dose was higher; 76.5 ng day^{-1}, 319.2 ng day^{-1}, and 420.2 ng day^{-1} for infants, children, and adults, respectively. Exposure time is a parameter that influences the inhalation risk most strongly. The incremental lifetime cancer risk (ILCR) was calculated according to Equation (2) and it was 3.35×10^{-10} for infants, 4.56×10^{-9} for children, and 1.08×10^{-8} for adults in the cold period, but in the warm period the ILCR was 15 times lower than in the cold. According to US EPA [53], a one in a million chance of developing an additional human cancer over a lifetime (ILCR = 10^{-6}) is considered to be an acceptable level risk. A lifetime risk of one in a thousand (ILCR = 10^{-3}) was considered to be a serious health threat. The results determined in this study were much lower than the acceptable risk level of 10^{-6}. Higher ILCR values were found by Majewski et al. [20], who calculated the ILCR for students and lecturers at Gliwice and Warsaw University. They found that the ILCR of students exposed to PAHs bounded to the PM$_1$ particle ranged from 5.49×10^{-8} (Warsaw) to 1.43×10^{-7} (Gliwice). However, these results were also below the acceptable risk level of 1×10^{-6}.

3.5. Influence of Meteorological Conditions and Gaseous Pollutants on PAH Concentrations

Spearman's correlation matrix (Tables 6 and 7) showed that most of the PAHs measured in the cold and warm periods correlated well with each other ($p < 0.05$). The correlation between PAHs was very strong and coefficients ranged from 0.52 to 0.99 in the cold and from 0.81 to 0.99 in the warm period. These very high correlation coefficients suggested their shared sources and activities. Negative correlations were established between all of the PAHs and temperature and between PAHs and rainfall in both measuring periods (cold and warm). Humidity and pressure were negatively correlated with concentrations of PAHs in the warm period. During the cold part of the year, long temperature inversion periods may occur. Such weather conditions are characterized by high pressure and impaired air mixing, which favors pollutant accumulation. These stable atmospheric conditions combined with increased emissions are the basic origins of elevated concentrations of PAHs during the cold season.

The negative correlation with temperature can be explained with the ability of PAHs to easily vaporize from particle to gas phase [50,54]. A significant correlation was found in the cold period between CO, SO$_2$, and all PAHs, as was a positive correlation between NO$_2$ and PAHs with high molecular weight, while a negative weak correlation but still statistically significant was established between heavy PAHs and O$_3$. Similar correlations were found during the warm period except between PAH and SO$_2$, where the correlation was not significant. As there is no industry at the location, SO$_2$ was mostly emitted from vehicles [55]. Previous studies have shown that SO$_2$ concentrations have been continuously decreasing over the last ten years, the reason for which is the fact that coal has not been in use for more than thirty years at this location [56]. Positive correlations between PAHs and NO$_2$ were in good agreement with other studies [55,57]. Previous studies have shown that NO$_2$ is one of the main traffic-related air pollutants, and in reaction with PAHs, it can produce nitro PAHs in the atmosphere [55]. In both measuring periods, a negative weak correlation but still significant was found between ozone and PAHs. Ozone is a reactive atmospheric pollutant, and it is able to, in the presence of heat and sunlight, react with PAHs so this was probably one of the reasons for the low PAHs concentrations during summer [55,58]. Positive correlations between CO and PAH suggested their shared sources and activities in both the cold and warm period. Similar results were found in other studies [59].

Table 6. Spearman's correlation coefficients between PAHs, gaseous pollutants, and some meteorological parameters during the cold period.

	Pyr	BaA	Chry	BjF	BbF	BkF	BaP	DahA	BghiP	IP	CO	SO_2	NO_2	O_3	t	RH	P	Rainfall
Flu	0.985	0.648	0.712	0.619	0.656	0.651	0.556	0.564	0.515	0.637	0.425	0.473	0.106	0.227	−0.706	0.3	0.09	−0.167
Pyr		0.718	0.776	0.685	0.721	0.719	0.627	0.624	0.591	0.705	0.472	0.458	0.158	0.178	−0.703	0.32	0.089	−0.164
BaA			0.972	0.965	0.966	0.964	0.963	0.944	0.947	0.932	0.663	0.283	0.455	−0.265	−0.499	0.427	0.097	−0.114
Chry				0.968	0.971	0.971	0.924	0.889	0.918	0.956	0.723	0.411	0.5	−0.138	−0.598	0.427	0.043	−0.142
BjF					0.997	0.997	0.979	0.953	0.981	0.986	0.744	0.384	0.519	−0.269	−0.487	0.444	0.061	−0.139
BbF						0.998	0.979	0.956	0.978	0.985	0.735	0.396	0.478	−0.261	−0.511	0.453	0.091	−0.145
BkF							0.973	0.944	0.973	0.99	0.737	0.388	0.496	−0.253	−0.519	0.446	0.065	−0.145
BaP								0.985	0.99	0.949	0.682	0.304	0.442	−0.374	−0.393	0.444	0.121	−0.113
DahA									0.972	0.908	0.645	0.299	0.379	−0.391	−0.375	0.446	0.159	−0.111
BghiP										0.955	0.712	0.292	0.497	−0.389	−0.375	0.431	0.101	−0.122
IP											0.747	0.409	0.515	−0.216	−0.507	0.444	0.037	−0.145
CO												0.392	0.694	−0.326	−0.406	0.459	−0.035	−0.001
SO_2													0.178	0.265	−0.518	0.179	−0.012	−0.092
NO_2														−0.25	−0.023	0.098	−0.226	0.016
O_3															−0.289	−0.343	−0.322	−0.137
t																−0.275	−0.035	0.199
RH																	0.035	0.286
P																		−0.239

t—temperature, RH—relative humidity, and P—atmospheric pressure. Statistically significant correlation coefficients ($p < 0.05$) are underlined.

Table 7. Spearman's correlation coefficients between PAHs, gaseous pollutants, and some meteorological parameters during the warm period.

	Pyr	BaA	Chry	BjF	BbF	BkF	BaP	DahA	BghiP	IP	CO	SO$_2$	NO$_2$	O$_3$	t	RH	P	Rainfall
Flu	0.97	0.861	0.892	0.887	0.9	0.888	0.877	0.812	0.88	0.878	0.489	0.102	0.448	−0.396	−0.554	−0.013	−0.094	−0.227
Pyr		0.947	0.953	0.92	0.938	0.931	0.93	0.825	0.904	0.912	0.459	0.075	0.382	−0.328	−0.602	−0.072	−0.153	−0.214
BaA			0.983	0.928	0.934	0.938	0.955	0.844	0.895	0.913	0.457	0.109	0.277	−0.28	−0.634	−0.053	−0.227	−0.098
Chry				0.925	0.947	0.946	0.949	0.833	0.914	0.926	0.481	0.115	0.346	−0.325	−0.656	−0.026	−0.239	−0.059
BjF					0.985	0.988	0.989	0.966	0.979	0.983	0.525	0.12	0.337	−0.303	−0.56	−0.0004	−0.202	−0.12
BbF						0.999	0.986	0.927	0.989	0.991	0.531	0.123	0.376	−0.309	−0.578	−0.032	−0.226	−0.143
BkF							0.99	0.93	0.987	0.99	0.52	0.118	0.355	−0.297	−0.58	−0.025	−0.233	−0.131
BaP								0.934	0.971	0.977	0.499	0.115	0.306	−0.286	−0.598	−0.008	−0.215	−0.124
DahA									0.937	0.941	0.575	0.205	0.313	−0.338	−0.483	0.042	−0.151	−0.057
BghiP										0.996	0.535	0.13	0.371	−0.331	−0.573	−0.008	−0.215	−0.123
IP											0.548	0.135	0.367	−0.329	−0.592	−0.008	−0.207	−0.112
CO												0.719	0.513	−0.578	−0.262	0.221	−0.048	0.004
SO$_2$													0.023	−0.226	0.001	0.085	−0.052	0.04
NO$_2$														−0.613	−0.159	0.222	0.013	−0.029
O$_3$															0.45	−0.611	−0.159	−0.118
t																−0.331	−0.163	0.064
RH																	0.262	0.214
P																		−0.161

t—temperature, RH—relative humidity, and p—atmospheric pressure. Statistically significant correlation coefficients ($p < 0.05$) are underlined.

4. Conclusions

Measurements of PAHs in the PM_1 particle fraction at an urban location in Zagreb, Croatia, showed seasonal differences with higher mass concentrations of PAH in the cold than in the warm period. The annual mass concentration for BaP was 0.765 ng m^{-3}, which indicated that the value set by Directive 2004/107/EC for BaP of 1 ng m^{-3} has not been exceeded. During both measurement periods, the contributions of heavy PAHs, characteristic for vehicle emissions, were much higher than those of light PAHs. Results of diagnostic ratios and PCA showed that the main emission source of PAHs associated with PM_1 in this study area was engine combustion of diesel or gasoline during the warm period but that in the cold period emission from domestic heating is, in addition to diesel, the dominant source. The total carcinogenic potency was estimated using toxic equivalency factors. The average TCP was 0.14 ng m^{-3} and 2.14 ng m^{-3} for the warm and cold period, respectively. The incremental lifetime cancer risk (ILCR) was determined for three age groups of the population: infants (0–1 year), children (5–19 year), and adults (20–70 year) and was below the maximum acceptable level (1×10^{-6}), revealing that carcinogenic risk posed to the three age groups via inhalation is acceptable. However, more than ten times higher PAH values in the cold part of the year, as well as associated health risk, emphasize the need for regular monitoring of PAH levels in smaller particle fractions, such as PM_1. These data are a valuable tool in future plans and actions to control PAH sources and to improve the quality of life of urban populations.

Supplementary Materials: The following are available online at http://www.mdpi.com/1660-4601/17/24/9587/s1, Table S1: Detection (DL) and quantification limits (QL) and recoveries (R) of PAHs analyzed by high-performance liquid chromatography (HPLC) with a fluorescence detector, Table S2: Comparison of PAH mass concentrations in PM1 particle fraction obtained in this study with other studies, Table S3: Risk parameters for different age groups, Figure S1: Wind roses. Dependence of PAH concentrations (ng m^{-3}) on wind direction, average wind velocities (m s^{-1}) and wind direction frequencies (%) for cold and warm measuring periods, Figure S2: Distribution of incremental lifetime cancer risk for adults, children and infants, derived using Monte Carlo simulation in cold and warm period.

Author Contributions: Conceptualization, I.J. and G.P.; data curation, I.J. and G.P.; validation, S.D., I.B. and M.L.; formal analysis, I.J. and Z.S.Š.; writing—original draft, I.J.; writing—review and editing, I.J., G.P., R.G. and M.L. All authors have read and agreed to the published version of the manuscript.

Funding: This research received no external funding.

Acknowledgments: These measurements were conducted within the internal scientific project "Organic content of PM_1 particle fraction" funded by Institute for Medical Research and Occupational Health (PI: R. Godec).

Conflicts of Interest: The authors declare that they have no conflict of interest.

References

1. Ravindra, K.; Sokhi, R.; Grieken, R.V. Atmospheric polycyclic aromatic hydrocarbons: Source attribution, emission factors and regulation. *Atmos. Environ.* **2008**, *42*, 2895–2921. [CrossRef]
2. ATSDR (Agency for Toxic Substances and Disease Registry). *Toxicology Profile for Polycyclic Aromatic Hydrocarbons*; US Department of Health and Human Services: Atlanta, GA, USA, 1995. Available online: https://www.atsdr.cdc.gov/toxprofiles/tp69.pdf (accessed on 20 December 2020).
3. Ströher, G.L.; Poppi, N.R.; Raposo, J.L., Jr.; de Souza, J.B.G. Determination of polycyclic aromatic hydrocarbons by gas chromatography—Ion trap tandem mass spectrometry and source identifications by methods of diagnostic ratio in the ambient air of Campo Grande, Brazil. *Microchem. J.* **2007**, *86*, 112–118. [CrossRef]
4. Teixeira, E.C.; Agudelo-Castañeda, D.M.; Fachel, J.M.G.; Leal, K.A.; Garcia, K.O.; Wiegand, F. Source identification ans seasonal variation of polycyclic aromatic hydrocarbons associated with atmospheric fine and coarse particles in the Metropolitan Area of Porto Alegre, RS, Brazil. *Atmos. Res.* **2012**, *118*, 390–403. [CrossRef]
5. Tobiszewski, M.; Namieśnik, J. PAH diagnostic ratios for the identification of pollution emission sources. *Environ. Pollut.* **2012**, *162*, 110–119. [CrossRef]

6. Perico, A.; Gottardi, M.; Boddi, V.; Bavazzano, P.; Lanciotti, E. Assessment of exposure to Polycyclic Aromatic Hydrocarbons in police in Florence, Italy, throught personal air sampling and biological monitoring of the urinary metabolite 1-hydroxypyrene. *Arch. Environ. Health* **2001**, *56*, 506–512. [CrossRef]

7. Piccardo, M.T.; Stella, A.; Redaelli, A.; Minoia, C.; Valerio, F. Newsagents personal exposures to benzo(a)pyrene in Genoa, Italy. *Atmos. Environ.* **2003**, *37*, 603–613. [CrossRef]

8. Piccardo, M.T.; Stella, A.; Redaelli, A.; Balducci, D.; Coradeghini, R.; Minoia, C.; Valerio, F. Personal daily exposures to benzo(a)pyrene of taxi drivers in Genoa, Italy. *Sci. Total Environ.* **2004**, *330*, 39–45. [CrossRef]

9. Oliveira, M.; Slezakova, K.; Fernandes, A.; Teixeira, J.P.; Delerue-Matos, C.; Pereira, M.; Morais, S. Occupational exposure of firefighters to polycyclic aromatic hydrocarbons in non-fire work environments. *Sci. Total Environ.* **2017**, *592*, 277–287. [CrossRef]

10. Boada, L.D.; Henríquez-Hernández, L.A.; Navarro, P.; Zumbado, M.; Almeida-González, M.; Camacho, M.; Álvarez-León, E.E.; Valencia-Santana, J.A.; Luzardo, O.P. Exposure to polycyclic aromatic hydrocarbons (PAHs) and bladder cancer: Evaluation from a gene-environment perspective in a hospital-based case-control study in the Canary Islands (Spain). *Int. J. Occup. Environ. Health* **2015**, *21*, 23–30. [CrossRef]

11. Delgado-Saborit, J.M.; Stark, C.; Harrison, R.M. Carcinogenic potential, levels and sources of polycyclic aromatic hydrocarbon mixtures in indoor and outdoor environments and their implications for air quality standards. *Environ. Int.* **2011**, *37*, 383–392. [CrossRef]

12. Wickramasinghe, A.P.; Karunaratne, D.G.G.P.; Sivakanesan, R. PM10-bound polycyclic aromatic hydrocarbons: Concentrations, source characterization and estimating their risk in urban, suburban and rural areas in Kandy, Sri Lanka. *Atmos. Environ.* **2011**, *45*, 2642–2650. [CrossRef]

13. WHO Regional Office for Europe. *Air Quality Guidelines*, 2nd ed.; WHO: Copenhagen, Denmark, 2000.

14. Squizzato, S.; Masiol, M.; Agostini, C.; Visin, F.; Formenton, G.; Harrison, R.M.; Rampazzo, G. Factors, origin and sources affecting PM1concentrations and composition at an urban background site. *Atmos. Res.* **2016**, *180*, 262–273. [CrossRef]

15. Yang, M.; Guo, Y.M.; Bloom, M.S.; Dharmagee, S.C.; Morawska, L.; Heinrich, J.; Jalaludin, B.; Markevychd, I.; Knibbsf, L.D.; Lin, S.; et al. Is PM_1 similar to $PM_{2.5}$? A new insight into the association of PM_1 and $PM_{2.5}$ with children's lung function. *Environ. Int.* **2020**, *145*, 106092. [CrossRef] [PubMed]

16. Iakovides, M.; Stephanou, E.G.; Apostolaki, M.; Hadjicharalambous, M.; Evans, J.S.; Koutrakis, P.; Achilleos, S. Study of the occurrence of airborne Polycyclic Aromatic Hydrocarbons associated with respirable particles in two coastal cities at Eastern Mediterranean: Levels, source apportionment and potential risk for human health. *Atmos. Environ.* **2019**, *213*, 170–184. [CrossRef]

17. Chen, S.C.; Liao, C.M. Health risk assessment on human exposed to environmental polycyclic aromatic hydrocarbons pollution sources. *Sci. Total. Environ.* **2006**, *366*, 112–123. [CrossRef]

18. Liu, Y.; Qin, N.; Liang, W.; Chen, X.; Hou, R.; Kang, Y.; Guo, Q.; Cao, S.; Duan, X. Polycyclic aromatic hydrocarbon exposure of children in typical household coal combustion environments: Seasonal variations, sources and carcinogenic risks. *Int. J. Environ. Res. Public Health* **2020**, *17*, 6520. [CrossRef]

19. Jakovljević, I.; Pehnec, G.; Vađić, V.; Čačković, M.; Tomašić, V.; Doko Jelinić, J. Polycyclic aromatic hydrocarbons in PM10, PM2.5 and PM1 particle fraction in an urban area. *Air Qual. Atmos. Health* **2018**, *11*, 843–854. [CrossRef]

20. Majewski, G.; Widziewicz, K.; Rogula-Kozłowska, W.; Rogula-Kopiec, P.; Kociszewska, K.; Rozbicki, T.; Majder-Łopatka, M.; Niemczyk, M. PM Origin or Exposure Duration? Health Hazards from PM-bound Mercury and PM-bound PAHs among students and lecturers. *Int. J. Environ. Res. Public Health* **2018**, *15*, 316. [CrossRef]

21. Pateraki, S.; Fameli, K.M.; Assimakopoulos, V.; Bougiatioti, A.; Maggos, T.; Mihalopoulos, N. Levels, sources and health risk of PM2.5 and PM1-bound PAHs across the Greater Athens Area: The role of the type of environment and the meteorology. *Atmosphere (Basel)* **2019**, *10*, 622. [CrossRef]

22. Kozielska, B.; Rogula-Kozłowska, W.; Klejnowski, K. Seasonal variations in health hazard from polycyclic aromatic hydrocarbons bound to submicrometer particles at three characteristic site in the heavily polluted polish region. *Atmosphere* **2015**, *6*, 1–20. [CrossRef]

23. Lewandowska, A.U.; Staniszewska, M.; Witkowska, A.; Machuta, M.; Falkowska, L. Benzo(a)pyrene parallel measurements in PM1and PM2.5in the coastal zone of the Gulf of Gdansk (Baltic Sea) in the heating and non-heating seasons. *Environ. Sci. Pollut. Res.* **2018**, *25*, 19458–19469. [CrossRef]

24. Křůmal, K.; Mikuška, P.; Večeřa, Z. Polycyclic aromatic hydrocarbons and hopanes in PM_1 aerosols in urban areas. *Atmos. Environ.* **2013**, *67*, 27–37. [CrossRef]
25. Padoan, S.; Zappi, A.; Atam, T.; Melucci, D.; Gambaro, A.; Formenton, G.; Popovicheva, O.; Nguyen, D.L.; Schnelle-Kreis, J.; Zimmermann, R. Organic molecular markers and sources contributions in a polluted municipality of north—East Italy: Extended PCA-PMF statistical approach. *Environ. Res.* **2020**, *186*, 109587. [CrossRef] [PubMed]
26. Pehnec, G.; Jakovljević, I. Carcinogenic potency of airborne polycyclic aromatic hydrocarbons in relation to the particle fraction size. *Int. J. Environ. Res. Public Health* **2018**, *15*, 2485. [CrossRef]
27. Ojeda-Castillo, V.; López-López, A.; Hernández-Mena, L.; Murillo-Tovar, M.A.; Díaz-Torres, J.; Hernández-Paniagua, I.Y.; del Real-Olvera, J.; León-Becerril, E. Atmospheric distribution of PAHs and quinones in the gas and PM1 phases in the Guadalajara Metropolitan Area, Mexico: Sources and health risk. *Atmosphere (Basel)* **2018**, *9*, 137. [CrossRef]
28. Agudelo-Castañeda, D.M.; Teixeira, E.C. Seasonal changes, identification and source apportionment of PAH in PM10. *Atmos. Environ.* **2014**, *96*, 186–200. [CrossRef]
29. Jakovljević, I.; Pehnec, G.; Vadjić, V.; Šišović, A.; Davila, S.; Bešlić, I. Carcinogenic activity of polycyclic aromatic hydrocarbons bounded on particle fraction. *Environ. Sci. Pollut. Res.* **2015**, *22*, 15931–15940. [CrossRef]
30. Šišović, A.; Pehnec, G.; Jakovljević, I.; Šilović Hujić, M.; Vadjić, V.; Bešlić, I. Polycyclic aromatic hydrocarbons at different crossroads in Zagreb, Croatia. *Bull. Environ. Contam. Toxicol.* **2012**, *88*, 438–442. [CrossRef]
31. Fu, S.; Yang, Z.Z.; Li, K.; Xu, X.B. Spatial characteristics and major sources of polycyclic aromatic hydrocarbons from soil and respirable particulate matter in a Mega-City, Chin. *Bull. Environ. Contam. Toxicol.* **2010**, *85*, 15–21. [CrossRef]
32. Zhang, Z.; Huang, J.; Yu, G.; Hong, H. Occurrence of PAHs, PCBs and organochlorine pesticides in the Tonghui River of Beijing, China. *Environ. Pollut.* **2004**, *130*, 249–261. [CrossRef]
33. Yunker, M.B.; Macdonald, R.W.; Vingarzan, R.; Mitchell, R.H.; Goyette, D.; Sylvestre, S. PAHs in the Fraser River basin: A critical appraisal of PAH ratios as indicators of PAH source and composition. *Org. Geochem.* **2002**, *33*, 489–515. [CrossRef]
34. Katsoyiannis, A.; Sweetman, A.J.; Jones, K.C. PAH Molecular diagnostic ratios applied to atmospheric sources: A critical evaluation using two decades of source inventory and air concentration data from the UK. *Environ. Sci. Technol.* **2011**, *45*, 8897–8906. [CrossRef] [PubMed]
35. Dvorská, A.; Lammel, G.; Klánová, J. Use of diagnostic ratio for studying source apportionment and reactivity of ambient polycyclic aromatic hydrocarbons over Central Europe. *Atmos. Environ.* **2011**, *45*, 420–427. [CrossRef]
36. Boström, C.; Gerde, P.; Hanberg, A.; Jernström, B.; Johansson, C.; Kyrklund, T.; Rannug, A.; Törnqvist, M.; Victorin, K.; Westerholm, R. Cancer risk assessment, indicators, and guidelines for polycyclic aromatic hydrocarbons in the ambient air. *Environ. Health Perspect.* **2002**, *110*, 451–488. [CrossRef] [PubMed]
37. Chu, M.; Chen, C. Evalution and estimation of potential carcinogenic risks of polynuclear aromatic hydrocarbons. In *Synposium on Polycyclic Aromatic Hydrocarbons in the Workplace*; Pacific Rim Risk Conference: Honolulu, HI, USA, 1984.
38. Clement. Comparative potency approach for estimating the cancer risk associated with exposure to mixture of polycyclic aromatic hydrocarbons (Interim Final Report). In *Prepared for EPA under Contract 68-02-4403*; ICF-Clement Associates: Fairfax, VA, USA, 1988.
39. Nisbet, I.C.T.; LaGoy, P.K. Toxic equivalency factors (TEFs) for polycyclic aromatic hydrocarbons (PAHs). *Regul. Toxicol. Pharmacol.* **1992**, *16*, 290–300. [CrossRef]
40. Rogula-Kozłowska, W.; Kozielska, B.; Klejnowski, K.; Szopa, S. Hazardous compounds in urban PM in the central part of Upper Silesia (Poland) in winter. *Arch. Environ. Prot.* **2013**, *39*, 53–65. [CrossRef]
41. Wenger, D.; Gerecke, A.C.; Heeb, N.V.; Hueglin, C.; Seiler, C.; Haag, R.; Naegeli, H.; Zenobi, R. Aryl hydrocarbon receptor-mediated activity of atmospheric particulate matter from an urban and a rural in Switzerland. *Atmos. Environ.* **2009**, *34*, 3556–3562. [CrossRef]
42. Chen, F.; Hu, W.; Zhong, Q. Emissions of particle-phase polycyclic aromatic hydrocarbons (PAHs) in the bFu Gui-shan Tunnel of Nanjing, China. *Atmos. Res.* **2013**, *124*, 53–60. [CrossRef]

43. Ho, K.F.; Ho, S.S.H.; Lee, S.C.; Cheng, Y.; Chow, J.C.; Watson, J.G.; Louie, P.K.K.; Tian, L. Emissions of gas and particle-phase polycyclic aromatic hydrocarbons (PAHs) in the Shing Mun Tunnel, Hong Kong. *Atmos. Environ.* **2009**, *43*, 6343–6351. [CrossRef]

44. Fang, G.; Chang, C.; Wu, Y.; Fu, P.; Yang, L.; Chen, M. Characterization, identification of ambient air and road dust polycyclic aromatic hydrocarbons in Central Taiwan Taichung. *Sci. Total. Environ.* **2004**, *327*, 135–146. [CrossRef]

45. Ravindra, K.; Bencs, L.; Wauters, E.; de Hoog, J.; Deutsch, F.; Roekens, E.; Bleux, N.; Berghmans, P.; Van Grieken, R. Seasonal and site-specific variation in vapour and aerosol phase PAHs over Flanders (Belgium) and their relation with anthropogenic activities. *Atmos. Environ.* **2006**, *40*, 771–785. [CrossRef]

46. Hanedar, A.; Alp, K.; Kaynak, B.; Avşar, E. Toxicity evaluation and source apportionment of Polycyclic Aromatic Hydrocarbons (PAHs) at three stations in Istanbil, Turkey. *Sci. Total. Environ.* **2014**, *488–489*, 437–446. [CrossRef] [PubMed]

47. Pehnec, G.; Jakovljević, I.; Godec, R.; Sever Štrukil, Z.; Žero, S.; Huremović, J.; Džepina, K. Carcinogenic organic content of particulate matter at urban locations with different pollution sources. *Sci. Total. Environ.* **2020**, *734*, 139414. [CrossRef] [PubMed]

48. Chang, K.F.; Fang, G.C.; Chen, J.C.; Wu, Y.S. Atmospheric polycyclic aromatic hydrocarbons (PAHs) in Asia: A review from 1999 to 2004. *Environ. Pollut.* **2006**, *142*, 388–396. [CrossRef]

49. Kulkarni, P.; Venkataraman, C. Atmospheric polycyclic aromatic hydrocarbons in Mumbai, India. *Atmos. Environ.* **2000**, *34*, 2785–2790. [CrossRef]

50. Galarneau, E. Source specificity and atmospheric processing of airborne PAHs: Implications for source apportionment. *Atmos. Environ.* **2008**, *42*, 8139–8149. [CrossRef]

51. Rogula-Kozłowska, W.; Kozielska, B.; Klejnowski, K. Concentration, origin and health hazard from fine particle-bound PAH at three characteristic sites in Southern Poland. *Bull. Environ. Contam. Toxicol.* **2013**, *91*, 349–355. [CrossRef]

52. Cvetković, A.; Jovašević-Stojanović, M.; Marković, D.; Ristovski, Z. Concentration and source identification of polycyclic aromatic hydrocarbons in the metropolitan area of Belgrade, Serbia. *Atmos. Environ.* **2015**, *112*, 335–343. [CrossRef]

53. US EPA (United States Environmental Protection Agency). *Guidance for Superfund Volume I: Human Health Evaluation Manual*; Part F, Supplemental Guidance for Inhalation Risk Assessment; US EPA: Washington, DC, USA, 2009; p. 68.

54. Callén, M.S.; López, J.M.; Mastral, A.M. Seasonal variation of benzo(a)pyrene in the Spanish airborne PM10. Multivariate linear regression model applied to estimate BaP concentrations. *J. Hazard Mater.* **2010**, *180*, 648–655. [CrossRef]

55. Tham, Y.W.F.; Takeda, K.; Sakugawa, H. Exploing the correlation of particulate PAHs, sulfor dioxide, nitrogen dioxide and ozone, a preliminary study. *Water Air Soil Pollut.* **2008**, *194*, 5–12. [CrossRef]

56. Pehnec, G. Trends of air quality in Zagreb. In Proceedings of the Regional Symposium on air Quality in Cities, Sarajevo, Bosna and Herzegovina, 30–31 January 2020. (In Croatian).

57. Park, S.S.; Kim, Y.J.; Kang, C.H. Atmospheric polycyclic aromatic hydrocarbons in Seoul, Korea. *Atmos. Environ.* **2002**, *36*, 2917–2924. [CrossRef]

58. Pudasainee, D.; Sapkota, B.; Shrestha, M.L.; Kaga, A.; Kondo, A.; Inoue, Y. Ground level ozone concentrations and its association with NO_x and meterological parameters in Kathmandu valley, Nepal. *Chemosphere* **2006**, *15*, 675–685. [CrossRef]

59. Masiol, M.; Hofer, A.; Squizzato, S.; Piazza, R.; Rampazzo, G.; Pavoni, B. Carcinogenic and mutagenic risk associated to airborne particle-phase polycyclic aromatic hydrocarbons: A source apportionment. *Atmos. Environ.* **2012**, *60*, 375–382. [CrossRef]

Publisher's Note: MDPI stays neutral with regard to jurisdictional claims in published maps and institutional affiliations.

International Journal of
**Environmental Research
and Public Health**

Article

Statistical Forecast of Pollution Episodes in Macao during National Holiday and COVID-19

Man Tat Lei [1,2,*], Joana Monjardino [3], Luisa Mendes [1], David Gonçalves [2] and Francisco Ferreira [3]

[1] Department of Sciences and Environmental Engineering, NOVA School of Science and Technology, NOVA University Lisbon, 2829-516 Caparica, Portugal; lc.mendes@fct.unl.pt

[2] Institute of Science and Environment, University of Saint Joseph, Macau 999078, China; david.goncalves@usj.edu.mo

[3] Center for Environmental and Sustainability Research, NOVA School of Science and Technology, NOVA University Lisbon, 2829-516 Caparica, Portugal; jvm@fct.unl.pt (J.M.); ff@fct.unl.pt (F.F.)

* Correspondence: l.tat@campus.fct.unl.pt or lei.man.tat@usj.edu.mo

Received: 27 May 2020; Accepted: 14 July 2020; Published: 15 July 2020

Abstract: Statistical methods such as multiple linear regression (MLR) and classification and regression tree (CART) analysis were used to build prediction models for the levels of pollutant concentrations in Macao using meteorological and air quality historical data to three periods: (i) from 2013 to 2016, (ii) from 2015 to 2018, and (iii) from 2013 to 2018. The variables retained by the models were identical for nitrogen dioxide (NO_2), particulate matter (PM_{10}), $PM_{2.5}$, but not for ozone (O_3) Air pollution data from 2019 was used for validation purposes. The model for the 2013 to 2018 period was the one that performed best in prediction of the next-day concentrations levels in 2019, with high coefficient of determination (R^2), between predicted and observed daily average concentrations (between 0.78 and 0.89 for all pollutants), and low root mean square error (RMSE), mean absolute error (MAE), and biases (BIAS). To understand if the prediction model was robust to extreme variations in pollutants concentration, a test was performed under the circumstances of a high pollution episode for $PM_{2.5}$ and O_3 during 2019, and the low pollution episode during the period of implementation of the preventive measures for COVID-19 pandemic. Regarding the high pollution episode, the period of the Chinese National Holiday of 2019 was selected, in which high concentration levels were identified for $PM_{2.5}$ and O_3, with peaks of daily concentration exceeding 55 $\mu g/m^3$ and 400 $\mu g/m^3$, respectively. The 2013 to 2018 model successfully predicted this high pollution episode with high coefficients of determination (of 0.92 for $PM_{2.5}$ and 0.82 for O_3). The low pollution episode for $PM_{2.5}$ and O_3 was identified during the 2020 COVID-19 pandemic period, with a low record of daily concentration for $PM_{2.5}$ levels at 2 $\mu g/m^3$ and O_3 levels at 50 $\mu g/m^3$, respectively. The 2013 to 2018 model successfully predicted the low pollution episode for $PM_{2.5}$ and O_3 with a high coefficient of determination (0.86 and 0.84, respectively). Overall, the results demonstrate that the statistical forecast model is robust and able to correctly reproduce extreme air pollution events of both high and low concentration levels.

Keywords: air pollution; air quality forecast; modelling; pollution episodes; national holiday; COVID-19

1. Introduction

The development of air quality forecast models is essential for cities with high population density, including Macao, one of the most densely populated cities in the world. It is extremely important to predict pollution episodes so the authority can provide a warning to the local community in advance to avoid the adverse air quality, which may lead to severe health consequences. In order to predict next-day concentrations of nitrogen dioxide (NO_2), particulate matter (PM_{10} and $PM_{2.5}$), and maximum hourly concentration of ozone ($O_{3\,MAX}$) for roadside, ambient, and residential stations in Macao, a forecast model was developed based on statistical methods using multiple linear regression (MLR) and classification and regression tree (CART) analysis.

There are three forms of total suspended particles (TSPs), which include coarse, fine, and ultrafine particles. Coarse particles, also known as PM_{10}, are derived from suspension of dust, soil, sea salts, pollen, mold, and other crustal materials. Fine particles, also known as $PM_{2.5}$, are derived from emissions from combustion process, including vehicles powered by petrol and diesel, wood burning, coal burning, and other industrial processes. Ultrafine particles are derived from combustion related sources such as vehicle exhausts and atmospheric photochemical reactions [1].

O_3 is the most important index substance for photochemical smog, one of the major air pollutants [2]. The formation of ground-level O_3 heavily depends on the concentration levels of volatile organic compounds (VOCs) and nitrogen oxides (NO_x) and meteorological factors such as wind speed, insolation, and temperature. $PM_{2.5}$ and O_3 pollutants are known to cause the most damages to the human respiratory and cardiovascular system. A study for Terengganu State, Malaysia, showed that high levels of O_3 occurring under dry and warm conditions during the southwest monsoon, were higher in industrial areas, and were positively correlated with the maximum daily temperature [3].

The emission of NO_x is primarily emitted from transportation and combustion process, while the emission of VOCs is primarily emitted from road traffic and the use of products containing organic solvents [4,5].

The emission of NO_x and VOCs is responsible for the O_3 formation, in particular rural areas being NO_x-sensitive while urban areas being VOC-sensitive. Nevertheless, the greater NO_x emission reductions have contributed to a widespread shift in the O_3 production regime from NO_x-saturated (high-NO_x) to NO_x-sensitive (low-NO_x) in some urban areas, while O_3 production in rural areas is even more sensitive to NO_x.

TSPs are primary contributors to premature death worldwide, with over four million premature deaths being recorded due to exposure to high levels of ambient $PM_{2.5}$ [6–8]. $PM_{2.5}$ can penetrate deep into the lungs when being inhaled, which leads to both acute and chronic health issues [1,6]. NO_2 and TSPs are responsible for 412,000 and 71,000 premature death per year, respectively, in the European Union [9,10]. Moreover, previous studies show a strong correlation between short-term exposure to NO_2 and both the number of hospital outpatients with eye and adnexa diseases (EADs) [11] and the number of hospital admission due to cardiovascular diseases (CVD) [12]. The Chinese National Ambient Air Quality Standard (NAAQS) has set the threshold of PM_{10}, $PM_{2.5}$, and $O_{3\,MAX}$ concentration at 150 $\mu g/m^3$, 75 $\mu g/m^3$, and 160 $\mu g/m^3$, respectively, while the WHO Air Quality Guideline has set the same thresholds at 50 $\mu g/m^3$, 25 $\mu g/m^3$, and 100 $\mu g/m^3$, respectively. Compliance with the thresholds set by the WHO for $PM_{2.5}$ could improve life expectancy in China by 0.14 years [13] and ambient air pollution has caused at least 3.7 million deaths, with more than 25% of deaths in Southeast Asia [14,15].

Air pollution forecasting models can provide important information for populations to adopt mitigation measures during high pollution days. To be useful, these models should be robust to deal with extreme variations in pollution levels, in particular during high-pollution peak days. Factors leading to extreme variation in pollution levels are diverse and include both human activities and meteorological factors.

In a study for Beijing, China, the reduction of traffic flow and vehicle emissions in downtown areas during the Chinese National Holiday, reduced air pollution, while, in contrast, fireworks during the Chinese New Year Holiday had the opposite effect [16]. When highway tolls were being waived for passenger vehicles during the Chinese National Holiday across the entire nation of China, air pollution increased by 20% and visibility decreased by 1 km, causing economic losses due to negative health impacts estimated at RMB 0.95 billion [17]. Nevertheless, the Chinese National Holiday is known to be a golden week of tourism, in which the Chinese tourist flock to different tourist destinations around the world to celebrate the national holiday. Due to the vibrant casinos and entertainment industry and close proximity to mainland China, Macao is also one of the favorite destinations for Chinese tourists, so the influx of tourist during the period of Chinese National Holiday may lead to an increase of emissions in Macao.

Likewise, the recent COVID-19 crisis has had an extreme impact in air pollution levels. The Wuhan Health Commission has first reported cases of pneumonia linked to the Wuhan wet market in Hubei Province, China, back in December 2019 [18]. Preventive measures were implemented soon after that abruptly reduced industrial activities and transportation. Nevertheless, the levels of air pollutants, in particular of $PM_{2.5}$, remained severe in northern China throughout the end of January 2020 due to adverse meteorological conditions that have overwhelmed the benefits of emission reduction in transportation and industrial sectors [19].

Previous work showed that there is an increase in the level of O_3 concentrations and a decrease in the level of NO_2, PM_{10}, and $PM_{2.5}$ concentration during the period of COVID-19 pandemic lockdown in several cities of China, due to the significant reduction of transportation and industrial activities [4,5,20,21].

Several methodologies have been developed and applied to forecast air quality across the world, including deterministic, statistical, and machine learning methods [22–26]. Some studies showed that statistical models are more accurate and efficient compared to deterministic models, particularly in regions with complexed terrain [27–30] Moreover, prediction of NO_2, PM_{10}, $PM_{2.5}$, and $O_{3\,MAX}$ concentrations based on MLR and CART models have been successfully implemented in Macao, Bangkok, Changsha City, Beijing, Bilbao, and Pakistan [26,31–35].

In this context, it is relevant to develop a reliable methodology to forecast the concentration of air pollutants, which is presented and tested for a high pollution episode (associated with the Chinese National Holiday) and a low pollution episode (during COVID-19 preventive measures).

2. Materials and Methods

The air quality and meteorological variables that were considered to build all of the air quality statistical models were obtained from Macao Meteorological and Geophysical Bureau (SMG). The air quality data was gathered from the air quality monitoring network, namely for: Macao Roadside, Macao Residential, Taipa Ambient, Taipa Residential, and Coloane Ambient stations, which have a suitable historic dataset of surface air quality measurements for the levels of NO_2, PM_{10}, $PM_{2.5}$, and O_3 concentrations. These background stations (residential and ambient) can capture the regional contribution of PM_{10} and $PM_{2.5}$. There is a higher population and traffic density in Macao Roadside and Macao Residential, which are located in the main peninsula, in comparison to Taipa Ambient, Taipa Residential, and Coloane Ambient stations, which are located on the outlying islands.

Meteorological data was obtained from surface observations at SMG's Taipa Grande Meteorological Station, hourly observations from automatic weather stations, such as temperature, relative humidity, precipitation, average wind speed, and dew point temperature, as well as upper-air observations (from Hong Kong King's Park location) such as geopotential heights, thickness, stability, temperature, relative humidity, and dew point temperature at various altitudes. In the present work, statistical models such as multiple linear regression (MLR), and classification and regression tree (CART), are developed, based on historical measurements of meteorological and air quality variables. Table 1 presents all the variables considered as predictors in the MLR and CART forecast models, as shown in previous work [22]. The air quality variables considered included the levels of NO_2, PM_{10}, $PM_{2.5}$, and $O_{3\,MAX}$ concentration from 00:00 to 23:00 of the previous day, two days and three days ago, and from 16:00 of the previous day and 15:00 of today. The meteorological variables being considered included the upper-air observations from King's Park location, Hong Kong Observatory, surface observations and other variables from the monitoring network of Macao Meteorological and Geophysical Bureau (SMG).

Table 1. Variables considered as predictors in the multiple linear regression (MLR) and classification and regression tree (CART) models in all of the air quality forecast models.

Variable Type	Variable Name	Variable Description (Units)/ Observations
Air quality variables	NO_2, PM_{10}, $PM_{2.5}$	Average hourly concentration values ($\mu g/m^3$)
	$O_{3\,MAX}$	Maximum hourly concentration values ($\mu g/m^3$)
	16D#, 23D#	23D#: 24-h concentration averaging period between 00h and 23h 16D#: 24-h concentration averaging period between 16h of D1 and 15h of D0 eg: PM10_16D1, O3_MAX_23D1.
	D0, D1, D2, D3	D0: Forecast Day; D1: Previous Day (Forecast Day-1); D2: Forecast Day-2; and D3: Forecast Day-3.
Meteorological variables	Upper-air obs. *	
	H1000, H850, H700, H500	Geopotential Height at 1000 hPa, 850 hPa, 700 hPa, and 500 hPa (m)/Indicator of synoptic-scale weather pattern.
	TAR925, TAR850, TAR700	Air Temperature at 925 hPa, 850 hPa, and 700 hPa (°C)/Measure of strength and height of the subsidence inversion.
	HR925, HR850, HR700	Relative Humidity at 925 hPa, 850 hPa, and 700 hPa (%).
	TD925, TD850, TD700	Dew Point Temperature at 925 hPa, 850 hPa, and 700 hPa (°C).
	THI850, THI700, THI500	Thickness at 850 hPa, 700 hPa, and 500 hPa (m)/Related to the mean temperature in the layer.
	STB925, STB850, STB700	Stability at 925 hPa, 850 hPa, and 700 hPa (°C)/Indicator of atmospheric stability.
Surface observations	T_AIR_MX, T_AIR_MD, T_AIR_MN	Maximum, Average, and Minimum Air Temperature (°C)
	HRMX, HRMD, HRMN	Maximum, Average, and Minimum Relative Humidity (%)
	TD_MD	Average dew point temperature (ground level) (°C)
	RRTT	Precipitation (mm)/Associated with atmospheric washout
	VMED	Average wind speed (m/s)/Related to dispersion
Other variables	DD	Duration of the day: number of hours of sun per day (h)
	FF	Week-day indicator (flag): weekday = 0, weekend = 1

Meteorological variables: * Daily sounding at 12H (GMT+8) at King's Park Meteorological Station—Hong Kong Observatory.

In this study, meteorological and air quality variables for 2013 to 2016, 2015 to 2018, and 2013 to 2018 were used to build three separate forecasting models. The 2013 to 2016 model was constructed for the initial evaluation for the application of the statistical model to forecast air quality in Macao, while the 2015 to 2018 models and the 2013 to 2018 models are a follow-up, to determine if any improvement could be made with two additional years of data. The comparison of extended data ranging from 5 to 6 years are considered to be adequate lengths to test if there is any significant difference between the time series. Simultaneously, it would not be ideal to trace back too far with the time series, because regional emissions are constantly changing, and therefore the level of pollutants concentration may also be changing. The dataset from 2019 was the most recent dataset, which would be used for the model validation for all the models. This study is an empirical approach and also region-specific, which may also be chemical-regime dependent.

The final selected variables to predict the levels of $PM_{2.5}$ and O_3 concentration are common to different locations of Macao air quality monitoring stations. Some variables initially selected were rejected from the forecast models due to collinearity. The final objective is to obtain prediction models with the lowest number of variables, but with the maximum explained variance as translated by the coefficient of determination (R^2).

After selecting the best model, it was applied to forecast pollution levels during an extremely high pollution episode, and a low pollution period. The high and low pollution selected episodes were, respectively: (i) the period of Chinese National Holiday, a week before the Chinese National Holiday from September 23rd to 30th, 2019, and the week during the Chinese National Holiday from October 1st to 7th, and (ii) the preventive measures period of COVID-19, from February 5th to 20th, 2020.

The statistical model was built using IBM SPSS Statistics version 26 with MLR (stepwise) and CART methods [26,36]. SPSS is a statistical software that is applied to solve research problems through hypothesis testing and predictive analysis.

Model performance indicators were calculated, such as, coefficient of determination (R^2), root mean square error (RMSE), mean absolute error (MAE), and systematic error (BIAS).

3. Results and Discussion

3.1. Air Quality Forecast Models

The model performance indicators obtained for the 2013 to 2016 model and for the 2013 to 2018 model, validated with 2019 data, are listed in Tables 2 and 3, respectively. The models chosen to figure in Tables 2 and 3 are the ones that performed the worst and best 2019 validation results.

The results showed that the model for the 2013 to 2018 period was the one that performed best in predicting next-day concentrations levels in 2019, with high R^2 between predicted and observed daily average concentrations (between 0.78 and 0.89 for all pollutants) and low RMSE, MAE, and BIAS. The additional two years of data helped to improve the air quality forecasting model. Nevertheless, with the two other models (2013–2016 and 2015–2018) a significant R^2 (between 0.78 and 0.89 for all pollutants) was also obtained, but it translated into a less reliable air quality forecast.

Regarding model performance indicators obtained per pollutant and station, the majority of models show a good agreement and a similar R^2 range values (from 0.81 to 0.89), except for $O_{3\,MAX}$, which is more difficult to predict. MLR was used for all pollutants, while CART analysis was used in almost all the $O_{3\,MAX}$ models (Tables 2 and 3). This CART analysis complement was an approach to obtain improved results, mainly regarding a better prediction of high pollutant levels.

Table 4 presents the final model equations obtained for each pollutant, per air quality monitoring station, in the 2013 to 2018 model. Additionally, the final equations used to predict the levels of NO_2, PM_{10}, $PM_{2.5}$, and O_{3_MAX} concentrations are presented in Table 4.

Table 2. Model performance indicators for the 2013 to 2016 model validation with 2019 data.

Station	Pollutant	Model Performance Indicator				Model Built Using Only MLR or CART and MLR	
		R^2	RMSE	MAE	BIAS	MLR	CART
Macao Roadside	PM_{10}	0.88	8.6	5.8	1.8	✓	
	$PM_{2.5}$	0.86	5.4	3.7	1.5	✓	
	NO_2	0.89	8.0	5.9	0.4	✓	
Macao Residential	PM_{10}	0.89	8.8	5.9	−0.3	✓	
	$PM_{2.5}$	0.87	5.2	3.3	0.7	✓	
	NO_2	0.86	7.7	5.5	−0.4	✓	
	$O_{3\,MAX}$	0.85	23.2	14.0	0.0	✓	
Taipa Ambient	PM_{10}	0.88	7.9	5.4	1.7	✓	
	$PM_{2.5}$	0.86	5.1	3.6	1.6	✓	
	NO_2	0.87	6.1	4.2	0.9	✓	
	$O_{3\,MAX}$	0.86	24.4	14.8	−2.1	✓	✓
Taipa Residential	PM_{10}	0.87	8.0	5.2	0.1	✓	
	$PM_{2.5}$	0.88	5.7	3.5	−0.1	✓	
	NO_2	0.87	5.6	4.2	0.8	✓	
	$O_{3\,MAX}$	0.78	20.9	12.7	1.3	✓	✓
Coloane Ambient	PM_{10}	0.88	8.7	6.2	2.4	✓	
	PM_{25}	0.86	5.4	3.7	1.3	✓	
	NO_2	0.81	7.8	5.5	−0.2	✓	
	$O_{3\,MAX}$	0.79	24.7	15.9	−3.6	✓	✓

Table 3. Model performance indicators for the 2013 to 2018 model validation with 2019 data.

Station	Pollutant	Model Performance Indicator				Model Built Using Only MLR or CART and MLR	
		R^2	RMSE	MAE	BIAS	MLR	CART
Macao Roadside	PM_{10}	0.88	8.4	5.6	1.5	✓	
	$PM_{2.5}$	0.87	5.2	3.3	0.2	✓	
	NO_2	0.89	7.9	5.8	−0.1	✓	
Macao Residential	PM_{10}	0.89	8.8	5.9	−0.1	✓	
	$PM_{2.5}$	0.87	5.2	3.3	0.8	✓	
	NO_2	0.86	7.7	5.5	0.0	✓	
	$O_{3\,MAX}$	0.85	23.2	14.0	0.0	✓	
Taipa Ambient	PM_{10}	0.88	7.8	5.1	0.8	✓	
	$PM_{2.5}$	0.86	4.8	3.1	0.2	✓	
	NO_2	0.87	6.1	4.2	1.0	✓	
	$O_{3\,MAX}$	0.86	23.7	14.7	−1.6	✓	✓
Taipa Residential	PM_{10}	0.88	7.9	5.1	0.2	✓	
	$PM_{2.5}$	0.88	5.6	3.5	−0.1	✓	
	NO_2	0.87	5.6	4.1	0.6	✓	
	$O_{3\,MAX}$	0.78	20.9	12.7	1.3	✓	✓
Coloane Ambient	PM_{10}	0.89	8.3	5.7	1.2	✓	
	PM_{25}	0.86	5.3	3.6	1.0	✓	
	NO_2	0.81	7.8	5.5	−0.1	✓	
	$O_{3\,MAX}$	0.79	24.3	15.3	−3.0	✓	✓

Table 4. Variables and model equations for each pollutant per air quality monitoring station in the 2013 to 2018 model.

Station	Pollutant	Model Equations
Macao Roadside	NO$_2$	NO$_2$ = 0.897 × NO$_2$_16D1 + 0.011 × H850 − 0.151 × HRMN
	PM$_{10}$	PM$_{10}$ = 0.913 × PM$_{10}$_16D1 + 0.015 × H850 − 0.208 × HRMD
	PM$_{2.5}$	PM$_{2.5}$ = 0.943 × PM$_{25}$_16D1 + 0.006 × H850 − 0.091 × HRMD
Macao Residential	NO$_2$	NO$_2$ = 0.913 × NO$_2$_16D1 + 0.007 × H850 − 0.087 × HRMN
	PM$_{10}$	PM$_{10}$ = 0.896 × PM$_{10}$_16D1 + 0.016 × H850 − 0.224 × HRMD
	PM$_{2.5}$	PM$_{2.5}$ = 0.926 × PM$_{25}$_16D1 + 0.004 × H850 − 0.176 × TD_MD
	O$_3$ MAX	O$_3$ MAX = 1.089 × O$_3$_MAX_16D1 − 0.344 × O$_3$_MAX_23D1 − 1.303 × TD_MD + 1.437 × T_AIR_MX
Taipa Ambient	NO$_2$	NO$_2$ = 0.914 × NO$_2$_16D1 + 0.004 × H850 + 0.734 × STB925
	PM$_{10}$	PM$_{10}$ = 0.905 × PM$_{10}$_16D1 + 0.014 × H850 − 0.205 × HRMD
	PM$_{2.5}$	PM$_{2.5}$ = 0.928 × PM$_{25}$_16D1 + 0.006 × H850 − 0.093 × HRMD
	O$_3$ MAX	If [O$_3$ MAX_16D1] ≤ 105.50 O$_3$ MAX = 1.034 × O$_3$_max_16D1 − 0.214 × O$_3$_max_23D1 + 0.019 × H850 − 0.236 × HRMN If [O$_3$ MAX_16D1] =]105.50; 181.87] O$_3$ MAX = 0.994 × O$_3$_max_16D1 − 0.433 × O$_3$_max_23D1 + 0.051 × H850 − 0.529 × HRMN If [O$_3$ MAX_16D1] > 181.87 O$_3$ MAX = 1.006 × O$_3$_max_16D1 − 0.472 × O$_3$_max_23D1 + 0.12 × H850 − 2.025 × HRMN
Taipa Residential	NO$_2$	NO$_2$ = 0.859 × NO$_2$_16D1 + 0.007 × H850 − 0.271 × TD_MD
	PM$_{10}$	PM$_{10}$ = 0.902 × PM$_{10}$_16D1 + 0.015 × H850 − 0.204 × HRMD
	PM$_{2.5}$	PM$_{2.5}$ = 0.938 × PM$_{25}$_16D1 − 0.607 × TD_MD + 0.703 × TAR925
	O$_3$ MAX	If [O$_3$ MAX_16D1] ≤ 129.12 O$_3$ MAX = 1.028 × O$_3$_max_16D1 − 0.238 × O$_3$_max_23D1 + 0.019 × H850 − 0.216 × HRMN If [O$_3$ MAX_16D1] = [129.12; 207.10] O$_3$ MAX = 0.958 × O$_3$_max_16D1 − 0.381 × O$_3$_max_23D1 + 0.061 × H850 − 0.751 × HRMN If [O$_3$ MAX_16D1] > 207.10 O$_3$ MAX = 1.12 × O$_3$_max_16D1 − 0.5 × O$_3$_max_23D1 + 0.14 × H850 − 2.818 × HRMN
Coloane Ambient	NO$_2$	NO$_2$ = 0.931 × NO$_2$_16D1 − 0.503 × TD_MD + 0.628 × TAR925
	PM$_{10}$	PM$_{10}$ = 0.904 × PM$_{10}$_16D1 + 0.015 × H850 − 0.214 × HRMD
	PM$_{2.5}$	PM$_{2.5}$ = 0.927 × PM$_{25}$_16D1 + 0.005 × H850 − 0.069 × HRMN
	O$_3$ MAX	If [O$_3$ MAX_16D1] ≤ 116.20 O$_3$ MAX = 1.021 × O$_3$_max_16D1 − 0.233 × O$_3$_max_23D1 + 1.650 × T_AIR_MX − 1.392 × TD_MD If [O$_3$ MAX_16D1] =]116.20; 186.92] O$_3$ MAX = 0.831 × O$_3$_max_16D1 − 0.397 × O$_3$_max_23D1 + 4.929 × T_AIR_MX − 3.384 × TD_MD If [O$_3$ MAX_16D1] > 186.92 O$_3$ MAX = 0.921 × O$_3$_max_16D1 − 0.482 × O$_3$_max_23D1 + 8.868 × T_AIR_MX − 8.582 × TD_MD

3.2. Air Quality During the High Pollution Episode

Taipa Ambient is the representative background location for Macao, and was chosen to assess the background levels of PM$_{2.5}$ and O$_3$ during the extreme pollution episode.

The influx of tourists coming to Macao, in light of the Chinese National Holiday, contributed to an high pollution episode that occurred during late September and early October 2019, with peak daily levels of PM$_{2.5}$ concentration exceeding 55 µg/m^3 and O$_3$ MAX levels exceeding 400 µg/m^3, largely exceeding the threshold level recommended by the WHO.

The levels of $PM_{2.5}$ and $O_{3\,MAX}$ concentrations for Taipa Ambient during the Chinese National Holiday in 2019 (from September to November) are presented in Figures 1 and 2.

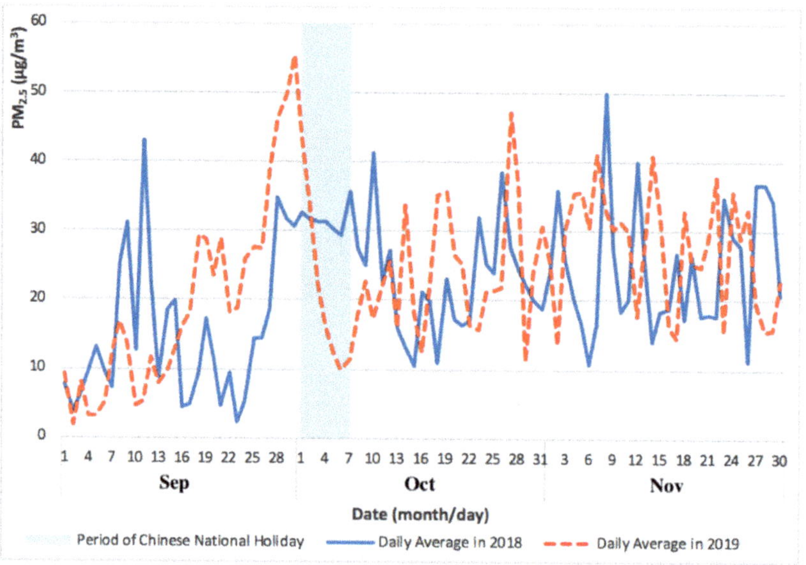

Figure 1. $PM_{2.5}$ concentrations for Taipa Ambient highlighting a pollution episode immediately before, and during, the Chinese National Holiday of 2018 and 2019 (September to November).

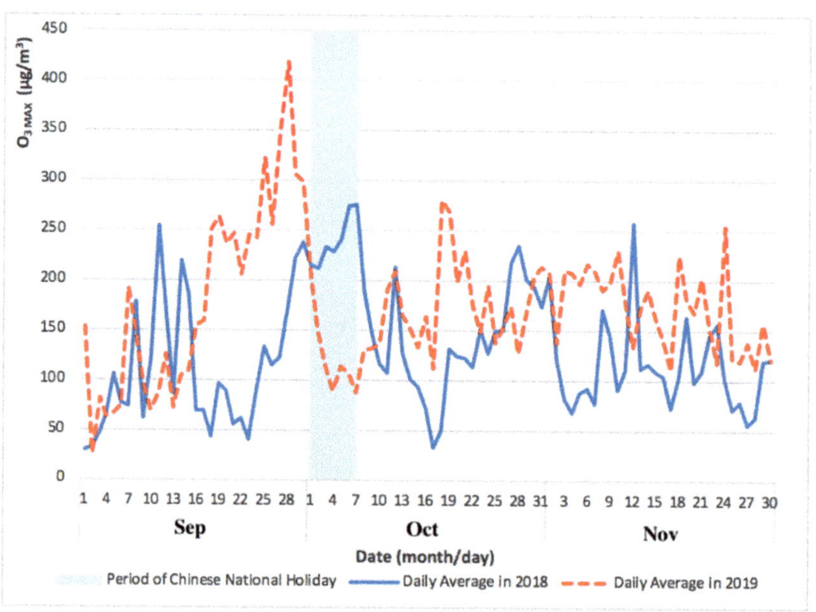

Figure 2. $O_{3\,MAX}$ concentrations for Taipa Ambient highlighting a pollution episode immediately before, and during, the Chinese National Holiday of 2018 and 2019 (September to November).

Figures 1 and 2 showed the comparison of daily average $PM_{2.5}$ and $O_{3\,MAX}$ concentration during 2018 and 2019, from a month before in September and a month after in November of the Chinese National Holiday. The pollution episode of 2019 occurred just before and going well into the period of Chinese National Holiday (1 to 7 October).

As shown in Figures 1 and 2, the levels of $PM_{2.5}$ and $O_{3\,MAX}$ concentration peaked immediately before, and during, the Chinese National Holiday in late September and early October 2019. The monthly mean concentration of $PM_{2.5}$ (from September to November) during the Chinese National Holiday in 2019 was 19 $\mu g/m^3$, 24 $\mu g/m^3$, and 28 $\mu g/m^3$, respectively. In addition, the monthly mean concentration of $O_{3\,MAX}$ (from September to November) during the Chinese National Holiday in 2019 was 181 $\mu g/m^3$, 163 $\mu g/m^3$, and 172 $\mu g/m^3$, respectively.

The levels of $O_{3\,MAX}$ concentrations reached its peak during the late September and early October due to meteorological factors including predominant winds from the north and east, from the Guangdong Province and Hong Kong, respectively. Temperatures were high in conjunction with low wind speed. The average daily temperature during the ozone peak episode that took place the two-weeks before the Chinese National Holiday (October 1st) was 28 °C, while the maximum daily average was 31 °C. Average wind speed was 2.5 m/s.

Due to the shutdown of nearby industrial sectors during the period of Chinese National Holiday, there were lower emissions of nitrogen oxides associated with the decreased load from the coal power plants in the northern region, usually supporting the operation of the factories. Therefore, this caused a decrease NO_x, the precursor of O_3. However, the increase in emissions of VOCs and NOx by vehicles, with chemical reactions in the presence of sunlight, may have caused the peak levels of ozone concentrations under these high temperature favorable conditions.

3.3. Air Quality During the Low Pollution Episode

In contrast, the COVID-19 pandemic has led to the Macao government's decision to temporarily suspend the operation of the casinos and entertainment industry and highly restrict cross border movements, as a preventive measure to reduce population mobility within the region of Macao. As a result, it has caused a low pollution episode during late January and early February 2020, with daily levels of $PM_{2.5}$ concentration reaching a record low at 2 $\mu g/m^3$ and $O_{3\,MAX}$ levels at 50 $\mu g/m^3$. The reduction of population mobility, and consequently, of traffic emissions in Macao and its nearby Guangdong Province, lead to this lowest $PM_{2.5}$ concentration levels.

As shown in Figure 3, the levels of $PM_{2.5}$ concentrations remained low during the initial outbreak of COVID-19 pandemic in Macao (from January to February 2020), slowly recovering to pre-COVID-19 values in March 2020. As shown in Figure 4, the levels of $O_{3\,MAX}$ concentration remained high during the initial outbreak of COVID-19 pandemic in Macao (from January to February 2020) and the high levels continued into March 2020. The higher levels of $O_{3\,MAX}$ concentration were associated with lower NO_x emissions, which led to a weakened O_3 titration by NO during the COVID-19 pandemic lockdown in the nearby Guangdong Province [4].

Despite industrial emission being a major contributor to the $PM_{2.5}$ pollution in China prior to COVID-19 pandemic lockdown period, the residential emission contributed to 39% of total $PM_{2.5}$ emissions in China, so the emissions of $PM_{2.5}$ during the lockdown period may have originated from residential areas [5].

The comparison of $PM_{2.5}$ and $O_{3\,MAX}$ concentrations for Taipa Ambient during the previous year of 2019 and COVID-19 pandemic in 2020 (January to March) is presented in Figures 3 and 4.

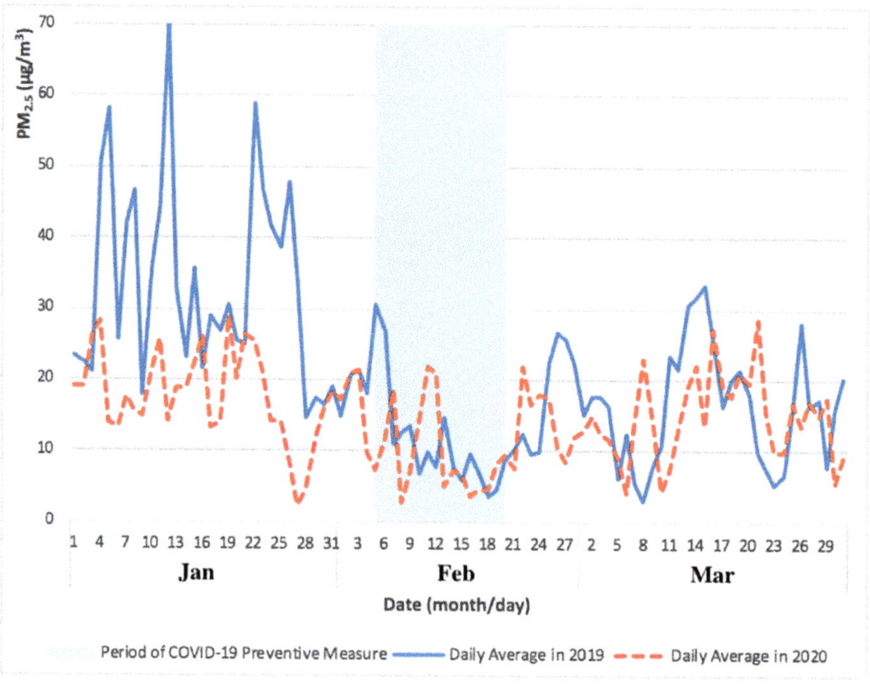

Figure 3. Comparison of $PM_{2.5}$ concentrations for Taipa Ambient during the previous year of 2019 and COVID-19 pandemic in 2020 (January to March).

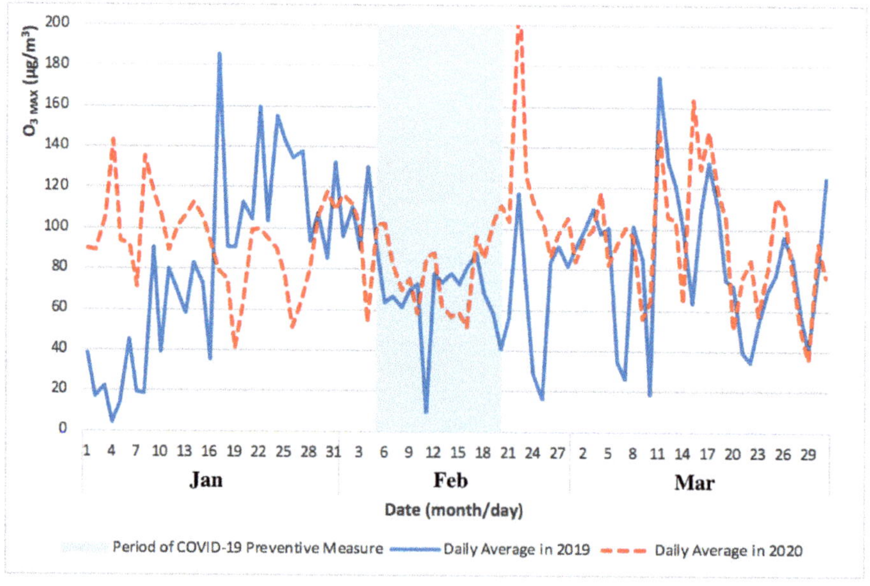

Figure 4. Comparison of $O_{3\,MAX}$ concentrations for Taipa Ambient during the previous year of 2019 and COVID-19 pandemic in 2020 (January to March).

As shown in Figure 5, the difference between monthly mean concentration (from January to March) of $PM_{2.5}$ concentration in 2019 and 2020 was 16 µg/m^3, 2 µg/m^3, and 1 µg/m^3, respectively. As shown in Figure 6, the difference between monthly mean concentration (from January to March) of $O_{3\,MAX}$ concentration in 2019 and 2020 was 12 µg/m^3, 21 µg/m^3, and 9 µg/m^3, respectively.

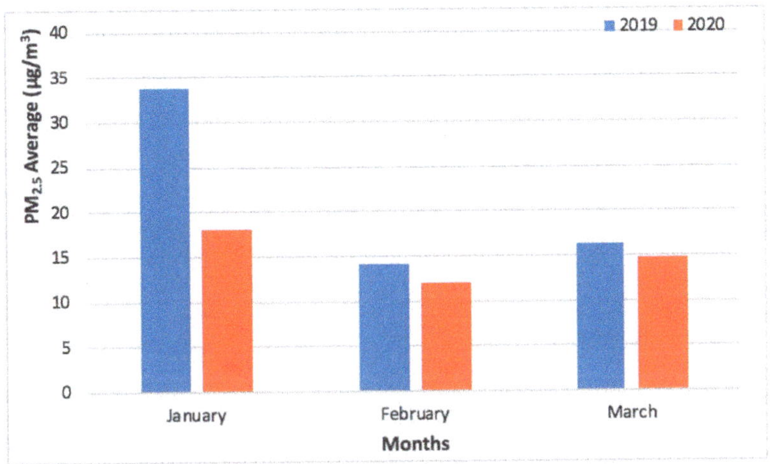

Figure 5. Monthly mean $PM_{2.5}$ concentrations for Taipa Ambient during the previous year of 2019 and COVID-19 pandemic in 2020 (January to March).

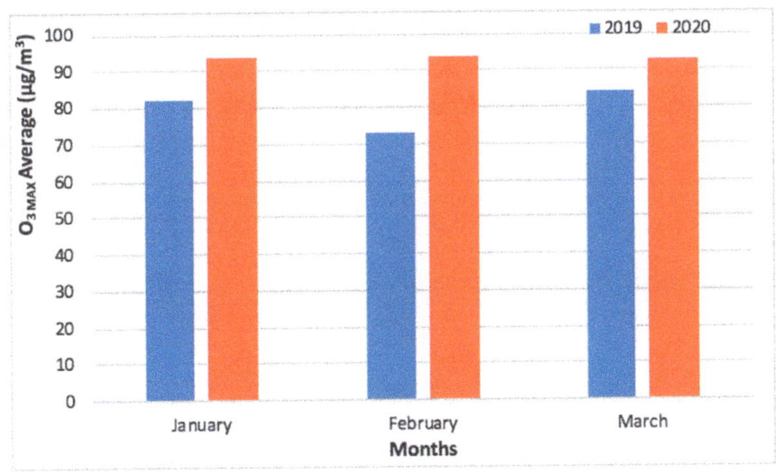

Figure 6. Monthly mean $O_{3\,MAX}$ concentrations for Taipa Ambient during the previous year of 2019 and COVID-19 pandemic in 2020 (January to March).

The monthly mean concentration of $PM_{2.5}$ and $O_{3\,MAX}$ concentration for Taipa Ambient during the previous year of 2019 and COVID-19 pandemic in 2020 (January to March) is presented in Figures 5 and 6. Overall, the preventive measures of COVID-19 pandemic may not have caused a significant difference in the levels of $PM_{2.5}$ and O_3 concentration in Macao, as the levels from February to March 2020 were similar to that of the previous year, 2019.

3.4. Air Quality Pollution Episodes Discussion

The air quality of Macao, a territory with only 32.8 km^2, is heavily influenced by external factors, in particular by human activities that occur in the much larger and neighboring Guangdong province. Our study shows the extent to which an increase in mobility associated with Chinese National Holiday, or a decrease in the same factors, associated with the COVID-19 preventive measures period, impacts air quality in Macao.

The levels of PM$_{2.5}$ concentrations significantly reduced after the first confirmed case of COVID-19 pandemic in Macao on January 22nd, 2020, which caused panic and anxiety in the local population, and continued by the announcement of casino closures by the Macao government as part of the preventive measures for COVID-19 from February 5th to 20th, 2020. As some of the preventive measures, in particular, the 15 days mandatory casino closure have been lifted, the fear and tension of the local residents has eased, which has promoted population mobility. Although the levels of PM$_{2.5}$ concentrations in Macao improved significantly during late January and early February 2020, the levels of PM$_{2.5}$ concentrations gradually returned to normal in March 2020 after some of the preventive measures began to be lifted in Macao and its nearby Guangdong Province.

3.5. Air Quality Pollution Episodes Forecast

Regarding the model behavior in predicting PM$_{2.5}$ and O$_{3\,MAX}$ during the high pollution episode (Chinese National Holiday), observed and predicted PM$_{2.5}$ and O$_{3\,MAX}$ concentrations are presented in Figures 7 and 8.

As shown in Figures 7 and 8, the levels of PM$_{2.5}$ and O$_{3\,MAX}$ concentration peaked during late September and early October 2019. The PM$_{2.5}$ predicted levels followed the primary trend of the measured concentrations and followed the concentration peak represented in Figure 7. The model for O$_{3\,MAX}$ also followed the primary trend, but it was more difficult to represent the concentration peak. The forecast model for PM$_{2.5}$ has a higher R^2 in comparison to the model of O$_{3\,MAX}$, because the maximum hourly concentration of O$_{3\,MAX}$ is more challenging to predict in comparison to the 24 h average of PM$_{2.5}$, as there is influence from the regional precursors sources and also its complex chemistry with solar radiation for O$_3$ formation, which led to a higher degree of variability.

Due to the different nature of PM$_{2.5}$ and O$_{3\,MAX}$, the forecast model performed better in the prediction of PM$_{2.5}$ in comparison to O$_{3\,MAX}$. This can be demonstrated in the higher R^2 values in the PM$_{2.5}$ forecast model. The observed and predicted PM$_{2.5}$ and O$_{3\,MAX}$ concentrations, during the low pollution episode (implementation of COVID-19 preventive measures), are presented in Figures 9 and 10.

The 2013 to 2018 model successfully predicted both the high and low pollution episodes, for PM$_{2.5}$ and O$_{3\,MAX}$, obtaining a significant R^2 of 0.88 and 0.83, respectively, for the high pollution period (from September to November 2019), and an R^2 of 0.82 and 0.75, respectively, for the low pollution period (from January to March 2020). The R^2 obtained for the entire year of 2019 was 0.86 for both PM$_{2.5}$ and O$_{3\,MAX}$. The statistical forecast model has been shown to be capable to predict, with a high coefficient of determination, the next 24 h.

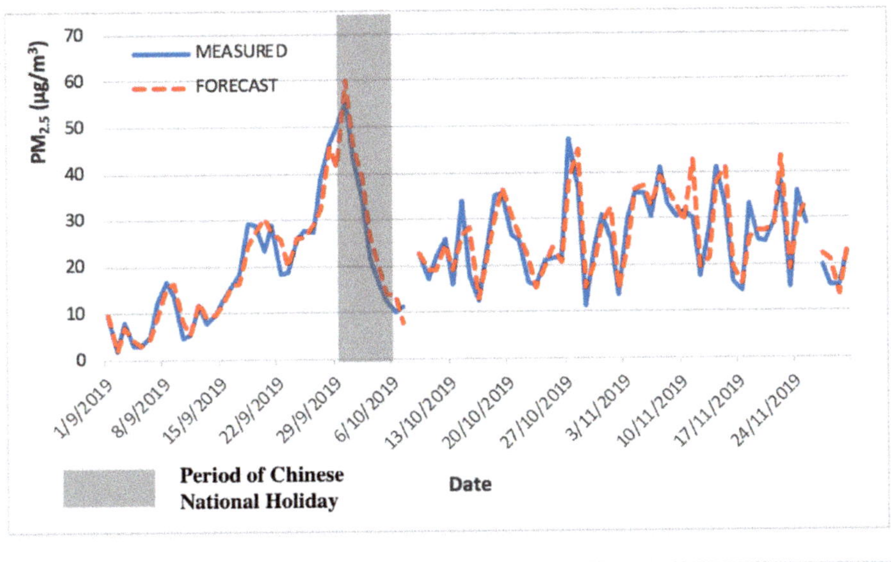

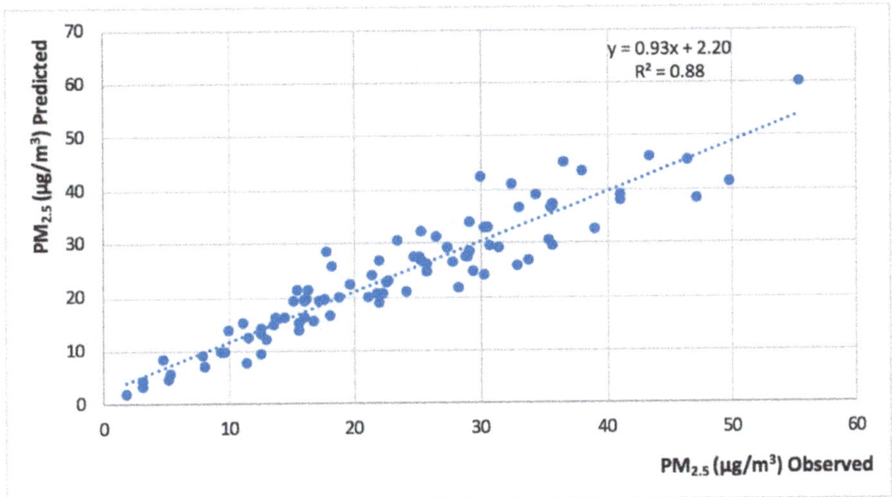

Figure 7. Observed and predicted $PM_{2.5}$ concentrations for Taipa Ambient during Chinese National Holiday (from September to November 2019).

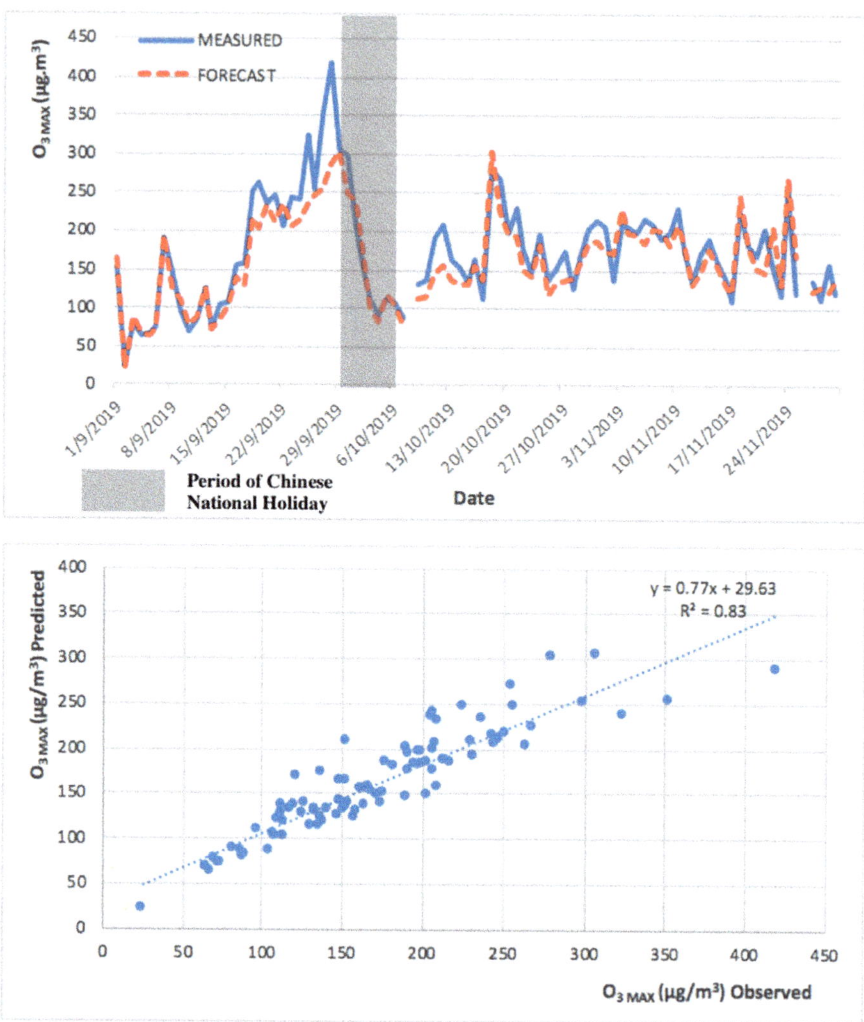

Figure 8. Observed and predicted $O_{3\,MAX}$ concentrations for Taipa Ambient during Chinese National Holiday (from September to November 2019).

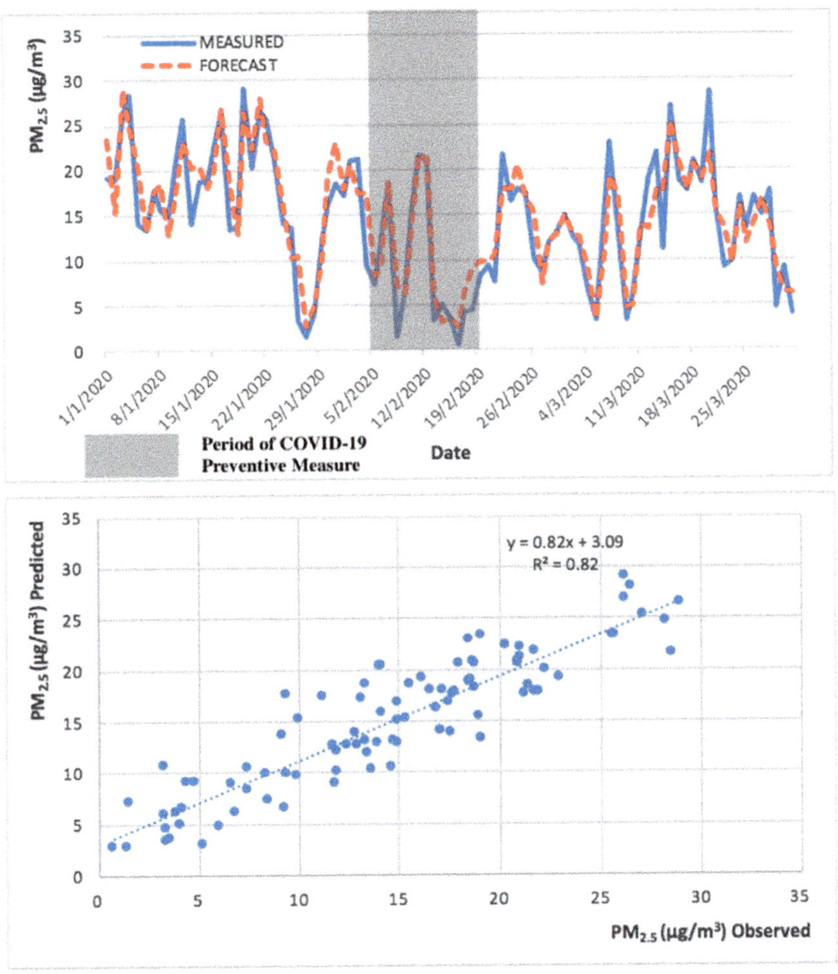

Figure 9. Observed and predicted PM$_{2.5}$ concentrations for Taipa Ambient during preventive measures of COVID-19 pandemic (from January to March 2020).

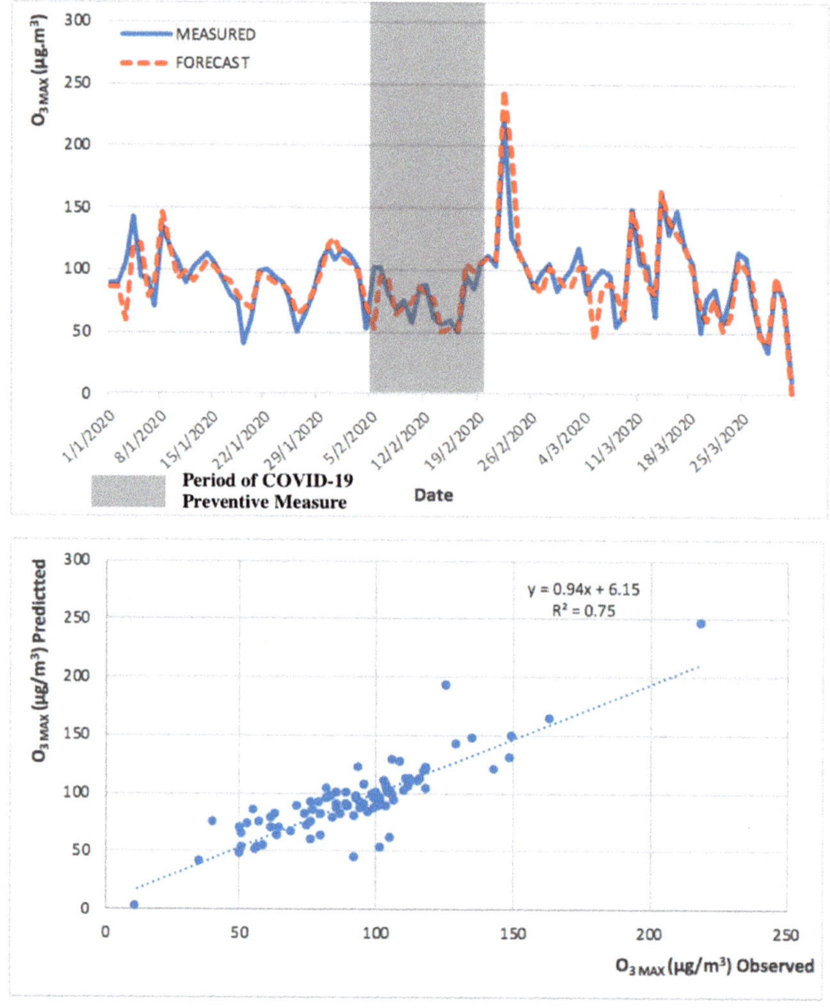

Figure 10. Observed and predicted $O_{3\,MAX}$ concentrations for Taipa Ambient during preventive measures of COVID-19 pandemic (from January to March 2020).

4. Conclusions

As expected, the 2013 to 2018 model performed best with the highest R^2 and lowest RMSE, MAE, and BIAS as compared with the 2013 to 2016 model and the 2015 to 2018 model. The additional two years of data helped to improve the accuracy and stability of the forecast of the 2013–2018 model.

The 2013–2018 model was able to successfully predict the high pollution episode during the Chinese National Holiday in late September and early October 2019 and the low pollution episode during the preventive measures period of COVID-19 pandemic in late January and early February 2020. This shows that this model can be reliably applied to forecast next-day pollutants concentrations across different magnitude levels of air pollution, being a useful tool for mitigation of air pollution impacts.

In addition, this shows that an improvement of global air quality in the territory is possible but it is tightly linked to the implementation of air pollution control measures in the industry and mobility sectors in Macao, in particular, in Guangdong Province. As previously studied, the air pollution problem associated with $PM_{2.5}$ and $O_{3\,MAX}$ is a regional problem that is not only limited to Macao, but also in the nearby regions of Hong Kong and Guangdong Province.

Author Contributions: Data curation, M.T.L., J.M., and L.M.; funding acquisition, D.G. and F.F.; methodology, M.T.L.; software, L.M.; supervision, D.G. and F.F.; validation, M.T.L., J.M., and L.M.; writing—original draft, M.T.L.; and writing—review and editing, M.T.L., J.M., D.G., and F.F. All authors have read and agreed to the published version of the manuscript.

Funding: This research was funded by [Fundação para a Ciência e Tecnologia, I.P., Portugal] grant number [UID/AMB/04085/2019] and the APC was funded by [CENSE].

Acknowledgments: The work developed was supported by The Macao Meteorological and Geophysical Bureau (SMG).

Conflicts of Interest: The authors declare no conflict of interest.

References

1. Pope, C.A.; Dockery, D.W. Health effects of fine particulate air pollution: Lines that connect. *J. Air Waste Manag. Assoc.* **2006**, *56*, 709–742. [CrossRef] [PubMed]
2. Ghazali, N.A.; Ramli, N.A.; Yahaya, A.S.; Yusof, N.F.F.M.; Sansuddin, N.; Al Madhoun, W.A. Transformation of nitrogen dioxide into ozone and prediction of ozone concentrations using multiple linear regression techniques. *Environ. Monit. Assess.* **2009**, *165*, 475–489. [CrossRef] [PubMed]
3. Abdullah, A.; Ismail, M.; Yuen, F.S.; Abdullah, S.; Elhadi, R. The relationship between daily maximum temperature and daily maximum ground level ozone concentration. *Pol. J. Environ. Stud.* **2017**, *26*, 517–523. [CrossRef]
4. Li, L.; Li, Q.; Huang, L.; Wang, Q.; Zhu, A.; Xu, J.; Liu, Z.; Li, H.; Shi, L.; Li, R.; et al. Air quality changes during the COVID-19 lockdown over the Yangtze River Delta Region: An insight into the impact of human activity pattern changes on air pollution variation. *Sci. Total Environ.* **2020**, *732*. [CrossRef]
5. Wang, Y.; Yuan, Y.; Wang, Q.; Liu, C.G.; Zhi, Q.; Cao, J. Changes in air quality related to the control of coronavirus in China: Implications for traffic and industrial emissions. *Sci. Total Environ.* **2020**, *731*, 139133. [CrossRef] [PubMed]
6. Wendt, E.A.; Quinn, C.W.; Miller-Lionberg, D.D.; Tryner, J.; Christian L'orange; Ford, B.; Yalin, A.P.; Pierce, J.R.; Jathar, S.H.; Volckens, J.; et al. A low-cost monitor for simultaneous measurement of fine particulate matter and aerosol optical depth-part 1: Specifications and testing. *Atmos. Meas. Tech.* **2019**, *12*, 5431–5441. [CrossRef]
7. Brauer, M.; Freedman, G.; Frostad, J.; Van Donkelaar, A.; Martin, R.V.; Dentener, F.; Van Dingenen, R.; Estep, K.; Amini, H.; Apte, J.; et al. Ambient air pollution exposure estimation for the global burden of disease 2013. *Environ. Sci. Technol.* **2015**, *50*, 79–88. [CrossRef]
8. Forouzanfar, M.H.; Afshin, A.; Alexander, L.T.; Anderson, H.R.; Bhutta, A.Z.; Biryukov, S.; Brauer, M.; Burnett, R.; Cercy, K.; Charlson, F.J.; et al. Global, regional, and national comparative risk assessment of 79 behavioural, environmental and occupational, and metabolic risks or clusters of risks, 1990–2015: A systematic analysis for the Global Burden of Disease Study 2015. *Lancet* **2016**, *388*, 1659–1724. [CrossRef]
9. EEA. *Air quality in Europe-2019 report-EEA Report No 10/2019*; European Environment Agency: Copenhagen, Denmark, 2019; ISBN 978-92-9480-088-6.
10. RCP. *Report of a working party February 2016*; Royal College of Physicians: London, UK, 2016; ISBN 9781860165672.
11. Song, J.; Liu, Y.; Lu, M.; An, Z.; Lu, J.; Chao, L.; Zheng, L.; Li, J.; Yao, S.; Wu, W.; et al. Short-term exposure to nitrogen dioxide pollution and the risk of eye and adnexa diseases in Xinxiang, China. *Atmos. Environ.* **2019**, *218*, 117001. [CrossRef]
12. Jevtic, M.; Dragić, N.; Bijelović, S.; Popović, M. Cardiovascular diseases and air pollution in Novi Sad, Serbia. *Int. J. Occup. Med. Environ. Health* **2014**, *27*, 153–164. [CrossRef]
13. Qi, J.; Ruan, Z.; Qian, Z.M.; Yin, P.; Yang, Y.; Acharya, B.K.; Wang, L.; Lin, H. Potential gains in life expectancy by attaining daily ambient fine particulate matter pollution standards in mainland China: A modeling study based on nationwide data. *PLoS Med.* **2020**, *17*, e1003027. [CrossRef] [PubMed]

14. Chen, W.; Li, A.; Zhang, F. Roadside atmospheric pollution: Still a serious environmental problem in Beijing, China. *Air Qual. Atmos. Health* **2018**, *11*, 1203–1216. [CrossRef]
15. Ai, Z.; Mak, C.M.; Lee, H. Roadside air quality and implications for control measures: A case study of Hong Kong. *Atmos. Environ.* **2016**, *137*, 6–16. [CrossRef]
16. Zheng, Y.; Che, H.; Zhao, T.; Zhao, H.; Gui, K.; Sun, T.; An, L.; Yu, J.; Liu, C.; Jiang, Y.; et al. Aerosol optical properties observation and its relationship to meteorological conditions and emission during the Chinese National Day and Spring Festival holiday in Beijing. *Atmos. Res.* **2017**, *197*, 188–200. [CrossRef]
17. Fu, S.; Gu, Y. Highway toll and air pollution: Evidence from Chinese cities. *J. Environ. Econ. Manag.* **2017**, *83*, 32–49. [CrossRef]
18. Ralph, R.; Lew, J.; Zeng, T.; Francis, M.; Xue, B.; Roux, M.; Ostadgavahi, A.T.; Rubino, S.; Dawe, N.J.; Al-Ahdal, M.N.; et al. 2019-nCoV (Wuhan virus), a novel Coronavirus: Human-to-human transmission, travel-related cases, and vaccine readiness. *J. Infect. Dev. Ctries.* **2020**, *14*, 3–17. [CrossRef]
19. Wang, P.; Chen, K.; Zhu, S.; Wang, P.; Zhang, H. Severe air pollution events not avoided by reduced anthropogenic activities during COVID-19 outbreak. *Resour. Conserv. Recycl.* **2020**, *158*, 104814. [CrossRef]
20. Shi, X.; Brasseur, G.P. The Response in Air Quality to the Reduction of Chinese Economic Activities During the COVID-19 Outbreak. *Geophys. Res. Lett.* **2020**, *47*, 1–8. [CrossRef]
21. Bao, R.; Zhang, A. Does lockdown reduce air pollution? Evidence from 44 cities in northern China. *Sci. Total Environ.* **2020**, *731*, 139052. [CrossRef]
22. Ma, J.; Cheng, J.C.; Lin, C.; Tan, Y.; Zhang, J. Improving air quality prediction accuracy at larger temporal resolutions using deep learning and transfer learning techniques. *Atmos. Environ.* **2019**, *214*, 116885. [CrossRef]
23. Masih, A. Machine learning algorithms in air quality modeling. *J. Environ Sci. Manag.* **2019**, *5*, 515–534. [CrossRef]
24. Lin, H.; Jin, J.; Herik, J.V.D. Air quality forecast through integrated data assimilation and machine learning. In Proceedings of the 11th International Conference on Agents and Artificial Intelligence, Prague, Czech Republic, 19–21 February 2019; Volume 2, pp. 787–793. [CrossRef]
25. Pagowski, M.; Grell, G.; Devenyi, D.; Peckham, S.; McKeen, S.; Gong, W.; Monache, L.D.; McHenry, J.; McQueen, J.; Lee, P. Application of dynamic linear regression to improve the skill of ensemble-based deterministic ozone forecasts. *Atmos. Environ.* **2006**, *40*, 3240–3250. [CrossRef]
26. Lei, M.T.; Monjardino, J.; Mendes, L.; Gonçalves, D.; Ferreira, F. Macao air quality forecast using statistical methods. *Air Qual. Atmos. Health* **2019**, *12*, 1049–1057. [CrossRef]
27. Kocijan, J.; Gradišar, D.; Stepančič, M.; Božnar, M.Z.; Grašič, B.; Mlakar, P. Selection of the data time interval for the prediction of maximum ozone concentrations. *Stoch. Environ. Res. Risk Assess.* **2017**, *32*, 1759–1770. [CrossRef]
28. Zhang, Y.; Bocquet, M.; Mallet, V.; Seigneur, C.; Baklanov, A. Real-time air quality forecasting, part I: History, techniques, and current status. *Atmos. Environ.* **2012**, *60*, 632–655. [CrossRef]
29. Zhang, Y.; Bocquet, M.; Mallet, V.; Seigneur, C.; Baklanov, A. Real-time air quality forecasting, part II: State of the science, current research needs, and future prospects. *Atmos. Environ.* **2012**, *60*, 656–676. [CrossRef]
30. Lopes, D.; Hoi, K.I.; Mok, K.M.; Miranda, A.I.; Yuen, K.V.; Borrego, C. Air quality in the main cities of the pearl river delta region. *Glob. Nest J.* **2016**, *18*, 794–802.
31. Chen, J.; Wang, J. Prediction of PM2.5 concentration based on multiple linear regression. In Proceedings of the 2019 International Conference on Smart Grid and Electrical Automation (ICSGEA), Xiangtan, China, 10–11 August 2019; pp. 457–460. [CrossRef]
32. Sahanavin, N.; Prueksasit, T.; Tantrakarnapa, K. Relationship between PM 10 and PM 2.5 levels in high-traffic area determined using path analysis and linear regression. *J. Environ. Sci.* **2018**, *69*, 105–114. [CrossRef]
33. Zhao, R.; Gu, X.; Xue, B.; Zhang, J.; Ren, W. Short period PM2.5 prediction based on multivariate linear regression model. *PLoS ONE* **2018**, *13*, e0201011. [CrossRef]
34. Samadianfard, S.; Delirhasannia, R.; Kişi, Ö.; Agirre-Basurko, E. Comparative analysis of ozone level prediction models using gene expression programming and multiple linear regression. *Geofizika* **2013**, *30*, 43–74.

35. Ahmad, M.; Alam, K.; Tariq, S.; Anwar, S.; Nasir, J.; Mansha, M. Estimating fine particulate concentration using a combined approach of linear regression and artificial neural network. *Atmos. Environ.* **2019**, *219*, 117050. [CrossRef]
36. Neto, J.; Torres, P.M.; Ferreira, F.; Boavida, F. Lisbon air quality forecast using statistical methods. *Int. J. Environ. Pollut.* **2009**, *39*, 333. [CrossRef]

International Journal of
Environmental Research and Public Health

Article

Five Year Trends of Particulate Matter Concentrations in Korean Regions (2015–2019): When to Ventilate?

Dohyeong Kim [1,†] , Hee-Eun Choi [2,†], Won-Mo Gal [2] and SungChul Seo [2,*]

[1] School of Economic, Political and Policy Sciences, The University of Texas at Dallas, 800 W Campbell Road, Richardson, TX 75080, USA; dohyeong.kim@utdallas.edu

[2] Department of Environmental Health and Safety, College of Health Industry, Eulji University, Seongnam 13135, Korea; heen001223@gmail.com (H.-E.C.); 1990134@eulji.ac.kr (W.-M.G.)

* Correspondence: seo@eulji.ac.kr; Tel.: +82-31-740-7143

† Joint first author.

Received: 6 June 2020; Accepted: 2 August 2020; Published: 10 August 2020

Abstract: Indoor air quality becomes more critical as people stay indoors longer, particularly children and the elderly who are vulnerable to air pollution. Natural ventilation has been recognized as the most economical and effective means of improving indoor air quality, but its benefit is questionable when the external air quality is unacceptable. Such risk-risk tradeoffs would require evidence-based guidelines for households and policymakers, but there is a lack of research that examines spatiotemporal long-term air quality trends, leaving us unclear on when to ventilate. This study aims to suggest the appropriate time for ventilation by analyzing the hourly and quarterly concentrations of particulate matter (PM)10 and PM2.5 in seven metropolitan cities and Jeju island in South Korea from January 2015 to September 2019. Both areas' PM levels decreased until 2018 and rebounded in 2019 but are consistently higher in spring and winter. Overall, the average concentrations of PM10 and PM2.5 peaked in the morning, declined in the afternoon, and rebounded in the evening, but the second peak was more pronounced for PM2.5. This study may suggest ventilation in the afternoon (2–6pm) instead of the morning or late evening, but substantial differences across the regions by season encourage intervention strategies tailored to regional characteristics.

Keywords: particulate matter; natural ventilation; indoor air quality; regional variation

1. Introduction

The quality of the air we breathe every day, both outdoors and indoors, is a matter of great concern since it contains various polluting substances which have negative effects on human health and on the environment [1]. According to the 2017 Global Burden of Disease Study, exposure to outdoor air pollution is one of the leading risk factors for premature death, accounting for 3.4 million deaths each year [2]. One of the most critical air pollutants is particulate matters (PM) which are considered as one of major reasons for increased prevalence or exacerbation of respiratory diseases [3,4], cardiovascular diseases [5], and diabetes [6]. The International Agency for Research on Cancer (IARC) has designated the atmospheric PM as a carcinogen of the same class as asbestos and those found in tobacco smoke [7]. However, a growing number of studies have reported that indoor PM could be more harmful in comparison to outdoor air quality because potential air pollutants are built up in confined environments [8,9]. Pollutants are endlessly generated in indoor environments due to human activities and inflow of external air and the purification of contaminated indoor air quality is difficult [10]. Due to a modern lifestyle in which people stay indoors for longer in most industrialized countries [11], the health impacts of indoor PM have been reported to outweigh those of outdoor air pollution, particularly for long-term exposure even at low concentration levels [12,13]. This is particularly relevant for susceptible groups such as children and elderly people who spend most of their time indoors [14,15].

Policymakers and professionals have developed strategies to manage indoor air quality such as improved ventilation and filtration, regulations and standard setting, routine monitoring and inspection, etc. [16]. Natural ventilation has been considered as the most economical and effective means of improving indoor air quality [17], which leads to improved productivity [18], mitigates allergic diseases [19], and even reduces the risk of airborne contagion [20]. Although natural ventilation may be less effective than mechanical ventilation in some settings [21], simply opening the windows for 10 min was found to be effective in achieving satisfactory indoor comfort conditions and air quality [22]. While natural ventilation is widely recommended for indoor air quality, its effectiveness depends on the quality of outdoor air [23,24]. Various concerns have been reported regarding its application, such as ambient particle concentration, indoor source intensity, indoor-outdoor air exchange rate, and circulation of outdoor air within the building, but the largest concern centers on the entry of polluted air from outdoors [25]. Due to the high filtration rate of particles from outdoors, the linkage between indoor and outdoor particle concentrations is more pronounced for a very small particle such as PM2.5 [26]. In particular, residential housing units are highly influenced by the quality of outdoor air since they tend to depend more on natural ventilation even if mechanical ventilation units are installed [27].

Considering a strong connectivity and dependency between indoor and outdoor air quality, natural ventilation could generate negative health impacts where the outdoor air quality is not acceptable for ventilating a building [28]. Despite the importance of risk-risk tradeoff between the exposure to indoor and outdoor air pollutants [29], except for a few governmental reports [17] there is no research that directly provides guidelines for the appropriate time and duration for natural ventilation based on the detailed assessment of temporal trends of ambient air pollutants in multiple communities. As hourly patterns of PM 2.5 and PM 10 could vary due to local characteristics such as major sources of contamination and seasonal weather conditions, an empirical investigation of long-term, site-specific air quality data is critical for developing evidence-based guidelines.

Despite the efforts of the Korean government to improve air quality during the past decades based on several laws and regulations such as the Clean Air Conservation Act of 1991 and the Special Act on the Improvement of Air Quality in Seoul Metropolitan Area of 2003, the problem of air pollution is far from being solved in South Korea. The public's concerns about the health risks associated with particulate matter exposure have substantially increased recently in South Korea, particularly after experiencing the record levels of outdoor PM 2.5 concentrations in many parts of the country in February and March 2019 [30]. Soon after these events, the national assembly of South Korea passed a series of bills which declared PM air pollution as a "social disaster" and provided emergency measures to tackle the problem, such as mandatory installation of air purifiers in classrooms, distribution of masks to vulnerable groups, and the promotion of low-emission vehicles [31]. It has been reported that Koreans spend more than 20 h a day indoors, particularly the most vulnerable groups such as patients, the elderly, and infants [32]. Although the Korea Ministry of Environment (KME) has provided several guidelines to citizens in response to indoor PM exposure, including ventilation, indoor water cleaning, and indoor air quality monitoring [33], they still lack details concerning how to actually implement the measures.

Only a few research articles have reported the seasonal patterns of particulate matter in South Korea, showing that PM10 and PM2.5 concentrations peak in spring and winter respectively, but both were lowest in summer [34]. However, to our knowledge there is no recent study in South Korea looking at hourly patterns of PM10 and PM2.5 concentrations by season in multiple regions based on long-term data. In South Korea, most people do not ventilate frequently due to Asian dust and air pollution problems attributed to industrial activities in China [35,36]. For this reason, indoor air quality of dwellings in South Korea often becomes worse due to anthropogenic activity including cooking which could lead to increases in the level of PM2.5 as well as NO_2 [37]. In this regard, KME suggested ventilation for over 30 min three times a day during daytime to improve indoor air quality [33], but this recommendation does not indicate a specific time and lacks tangible evidence of dynamic fluctuation patterns of each pollutant. Therefore, this study aims to analyze the historical records of atmospheric PM10 and PM2.5 concentrations in major cities and regions over the past five years (2015–2019) to assess hourly trends of each pollutant for each season and discuss potential reasons behind the patterns. The findings of this study should provide guidance on the most desirable and undesirable time during the day for not only natural ventilation but also outdoor activities, which could reduce the level of exposure to particulate matter and minimize the adverse effects of air pollution. Moreover, the improved understanding of dynamic trends of PM concentrations and their variations across multiple regions and seasons could help policymakers design a more effective strategy in identifying and monitoring region-specific sources of atmospheric air pollution.

The structure of this paper is as follows: the next section gives an overview of the eight study regions along with data collection and analytic processes. The following sections show the results of the analyses focusing on hourly trends of outdoor PM concentrations by year, season and region, followed by the discussion and conclusion sections that summarize the findings of this study and the implications and limitations of these findings.

2. Materials and Methods

The hourly PM10 and PM2.5 concentration data supplied by Air Korea (www.airkorea.or.kr) for KME were obtained for seven metropolitan cities (Seoul, Incheon, Daejeon, Daegu, Ulsan, Gwangju, Busan) and Jeju Island, for the past five years from January 2015 to September 2019. Figure 1 shows the mapped locations of the eight regions in South Korea and Table 1 summarizes the sociodemographic, geographic, meteorological, traffic and other relevant information for each region.

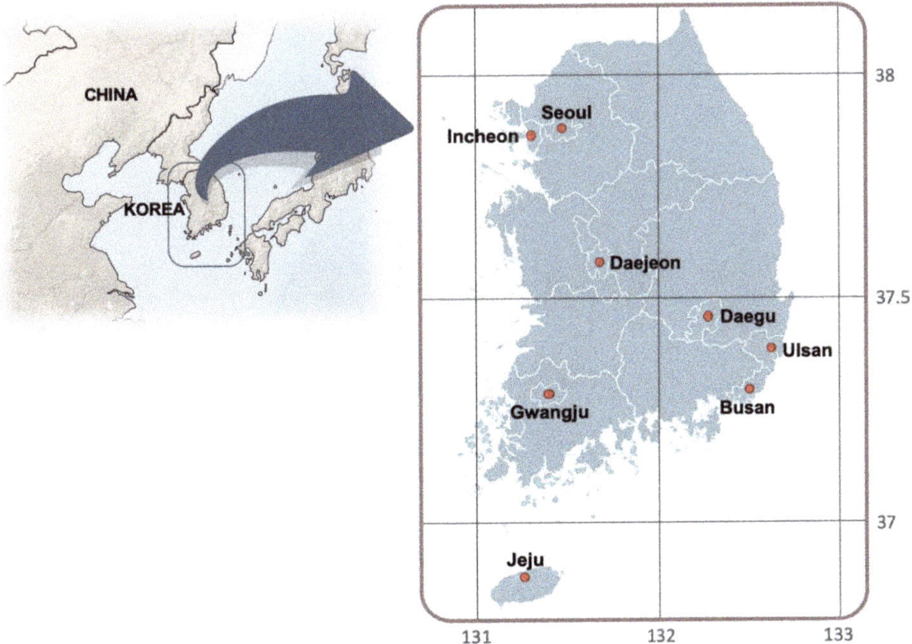

Figure 1. Location of the study regions in South Korea.

Table 1. Geographic, climatic, socio-economic and other notable characteristics in the eight regions.

Region	Area (km^2)	Population Density (people/km^2)	Average Temperature (°C)	Annual Average Precipitation (mm)	Number of Monitoring Stations	Notable Characteristics (Geographic, Meteorological and Mobility Patterns)
Seoul	605	15,964	13.5 (−10.9–36.8)	891.3	25	Capital of South Korea; located in the northwest of the country; heavy traffic congestion during rush hours
Busan	770	4380	15.7 (−4.4–35)	1623.2	25	Located in the south; relatively warm; close to the sea; heavy traffic congestion during rush hours
Daegu	883	2753	14.8 (−7.2–36.9)	995.7	15	Located inlands in south-central region; a basin-type city; high summer temperatures and frequent heat waves; heavy traffic congestion during rush hours
Incheon	1063	2769	13.2 (−10.4–36)	919.5	20	Located in the west; close to the sea; heavily influenced by the northwest wind; contains an industrial complex; heavy traffic congestion during rush hours
Gwangju	501	2980	14.7 (−5.9–34.8)	1085.9	9	Located in the southwest; heavily influenced by the northwest wind in winter; heavy traffic congestion during rush hours
Daejeon	540	2796	14 (−9.6–36)	984.2	10	Located around the center; a basin-type city; power plants around it; heavy traffic congestion during rush hours
Ulsan	1062	1080	14.9 (−5.3–35)	1045.1	17	Large-scale industrial complexes; close to the sea; distinct traffic patterns during rush hours
Jeju	1850	356	16.8 (1–35.4)	1979.9	6	An island city located in the southernmost part of Korea; a large number of tourists and no industrial facilities; constant traffic due to tourism

The observations from multiple monitoring stations in each region were integrated to calculate the average; very few observations were missing due to impediment or equipment failures during the period, and these were excluded from the analysis. Statistical analysis software (SAS for Windows version 9.1; SAS Institute Inc., Cary, NC, USA) was used to calculate the geometric means of PM10 and PM2.5 concentrations ($\mu g/m^3$) for each hour of a day (24 time slots) in each region, which were then aggregated by year and season to investigate the annual and seasonal changes and variations. Geometric means were chosen as the best summary statistic for the PM data since their distributions were highly skewed. Four distinct seasons in Korea were classified as: spring (March–May), summer (June–August), fall (September–November), and winter (December–February). A series of hourly time graphs were created to illustrate the trends of PM concentrations over a day for each year, season and region, in order to identify good and bad time slots for natural ventilation. Some additional analysis and discussion followed to explain the potential sources of the patterns, including hourly traffic volume in each region.

3. Results

3.1. Hourly Trends of Outdoor PM Concentrations by Year

Figure 2 shows the hourly patterns of PM10 and PM2.5 concentrations as an annual average for each year between 2015 and 2019. For both pollutants, there has been a gradual reduction from 2016 until 2018, but they rebounded in 2019. The PM10 concentrations were mostly below the KME's annual average standard of 50 $\mu g/m^3$ throughout a single day, while the PM2.5 concentrations were way above the recently-strengthened annual average standard of 15 $\mu g/m^3$ during the past five years. Despite some annual fluctuations, the overall hourly patterns for both PM10 and PM2.5 show some similarity throughout the five years; they peaked at 8–11am, declined in the afternoon and rose again in the evening. The second peak in the evening until dawn appears more conspicuous in PM2.5 concentrations, which is due to insufficient air circulation via inversion layer caused by the temperature drop on the earth's surface during these time ranges [38].

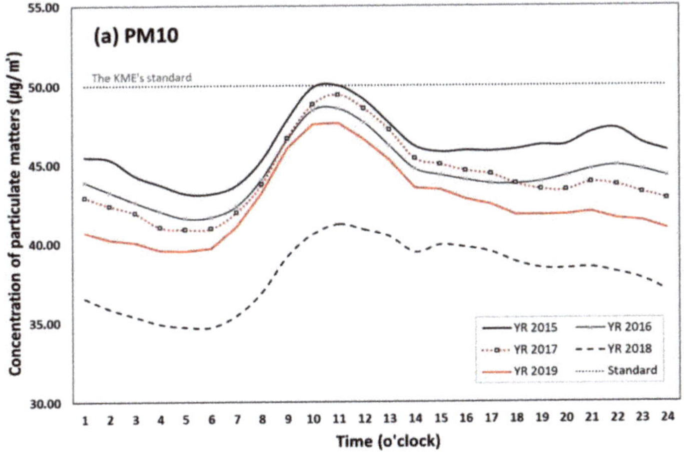

Figure 2. *Cont.*

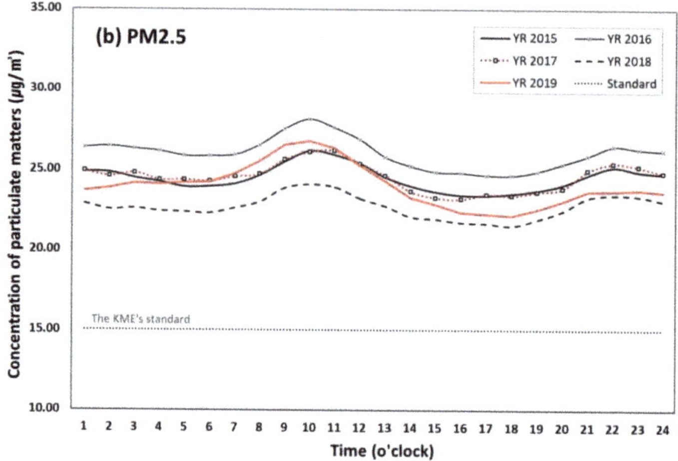

Figure 2. Diurnal variation of particulate matter (PM) concentrations (μg/m^3) by year (all regions).

3.2. Hourly Trends of Outdoor PM Concentrations by Season

Figure 3 compares the hourly patterns of both pollutants across the four seasons, showing a noticeable seasonal pattern: high in spring and winter and low in summer and fall. Similar to the previous study [34], PM10 concentration was highest in spring while PM2.5 concentration was highest in winter. Regardless of season, both PM10 and PM2.5 peaked between 9am–noon and decayed afterwards. The slope of decline in the afternoon was much steeper for PM2.5 than PM10, except for summer, when both concentration levels appear relatively stagnant throughout a day. The PM10 concentrations during spring and winter were above the KME's annual average standard of 50 μg/m^3 particularly between 8am and 2pm, while these were under the standard during summer and fall. PM2.5 concentrations were, however, much higher than the annual average standard of 15 μg/m^3 during all four seasons.

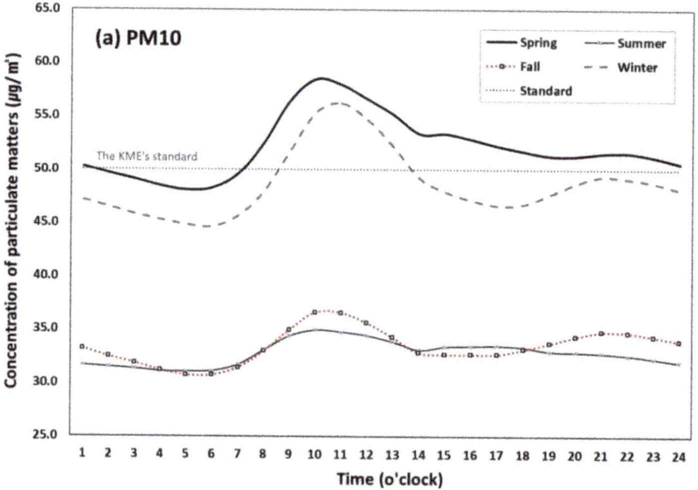

Figure 3. *Cont.*

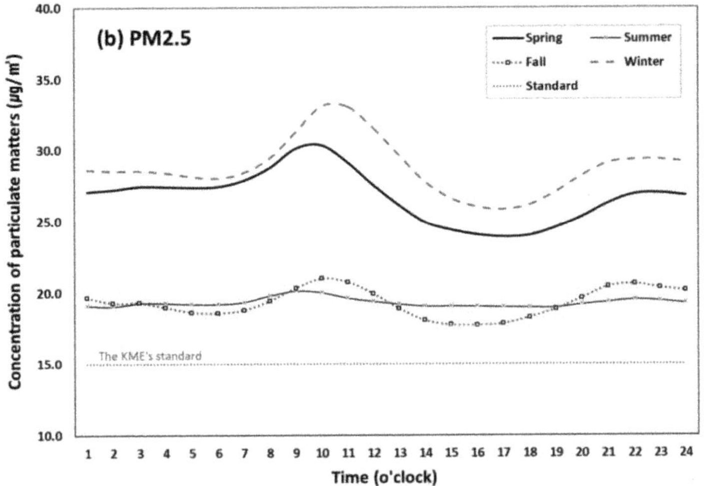

Figure 3. Diurnal variation of PM concentrations (µg/m³) by season (2015–2019).

3.3. Hourly Trends of Outdoor PM Concentrations by Season: Regional Variation

Figures 4 and 5 show regional variations of hourly and seasonal trends of PM10 and PM2.5 respectively, including Seoul (capital city of South Korea), six metropolitan cities (Incheon, Daejeon, Daegu, Ulsan, Gwangju, Busan) and Jeju Island. As seen in Figure 4, the seasonal and hourly patterns of PM10 are relatively similar across all regions, exhibiting high concentrations in spring and winter and a peak between 7am and 1pm, but the height of the peak and the declining patterns after the peak seem to vary by region. PM10 concentrations in Busan, Incheon and Ulsan were above the KME's annual average standard of 50 µg/m³ in the daytime period during spring, but its duration was much longer in Incheon (9am to 6pm) than Busan (9am–1pm) and Ulsan (10–11am). During winter, however, only Incheon and Daegu show PM10 concentration above the standard for a relatively short period (10am to noon). As for PM2.5, illustrated in Figure 5, the regional variation appears more noticeable. The PM2.5 concentrations are relatively similar between spring and winter in Seoul, Busan, Incheon, Gwangju, Ulsan and Jeju Island, while the winter concentrations were substantially larger than the spring concentrations in Daejeon and Daegu. For all eight regions, the spring and winter concentrations of PM2.5 were above the annual average standard of 15 µg/m³ throughout a given day. However, during the summer and fall seasons, the concentrations were above the standard only during the peak time periods in some regions. Unlike PM10, all regions except for Daegu show two peaks: one in the morning and the other in the evening or night.

The distinct patterns of PM10 and PM2.5 concentrations across the regions is mainly due to different size of particles. Relatively speaking, PM10 is associated with dust on roads, while PM2.5, a very tiny air particulate matter, even smaller than PM10, and relates to emission from vehicles such as aerosol and nitrogen oxide (NOx) [39]. For instance, the hourly trends of PM2.5 concentrations in Ulsan and Jeju Island look quite different from those of other regions, particularly in the afternoon and evening, possibly because of its unique traffic patterns. Ulsan is an industrial city known for factories and plumes leading to male-dominated demographics and workforce characteristics [40]. Thus, considering relatively less traffic congestion during commuting time in the city, air quality might be related more to industrial emissions than local traffic emissions. Jeju Island is a famous tourist destination located at the southern end of South Korea, exhibiting unique demographic, traffic and environmental characteristics. To further investigate the impact of traffic volume on PM2.5 concentration, the traffic volume data were obtained for the same eight regions for three years between 2016 and 2018 and the average amount of traffic was calculates for each hour. Figure 6 shows somewhat similar hourly patterns to the PM2.5 patterns found in Figure 5, beginning to increase at around 6am, fluctuating during the day, and declining at 6pm. This degree of similarity could indicate the level of contribution of traffic emissions on PM2.5 concentration trends, which should vary across the regions depending on site-specific characteristics [31,41]. As mentioned above, the traffic volumes in Ulsan and Jeju Island are substantially lower and flatter than the other regions due to their unique sociodemographic and mobility patterns.

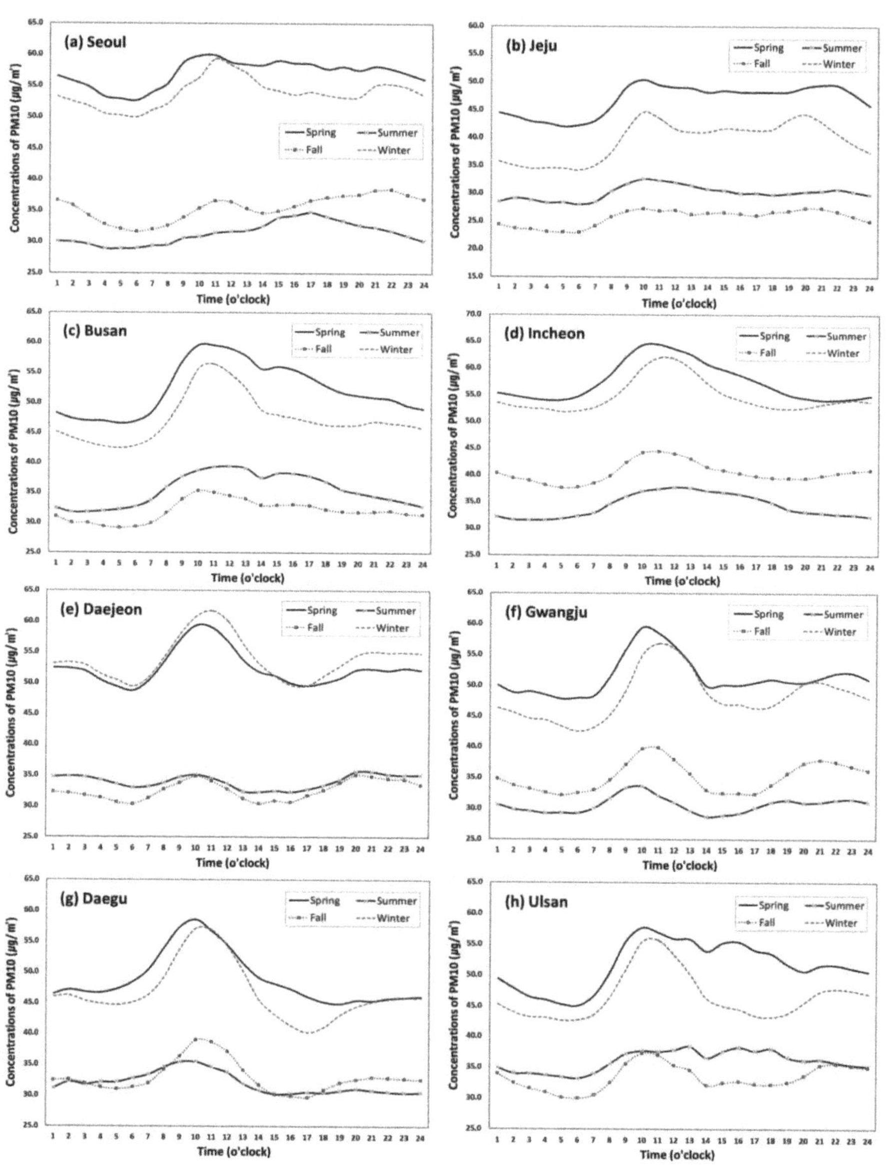

Figure 4. Diurnal variation of PM10 concentrations (μg/m^3) by season in each region (2015–2019).

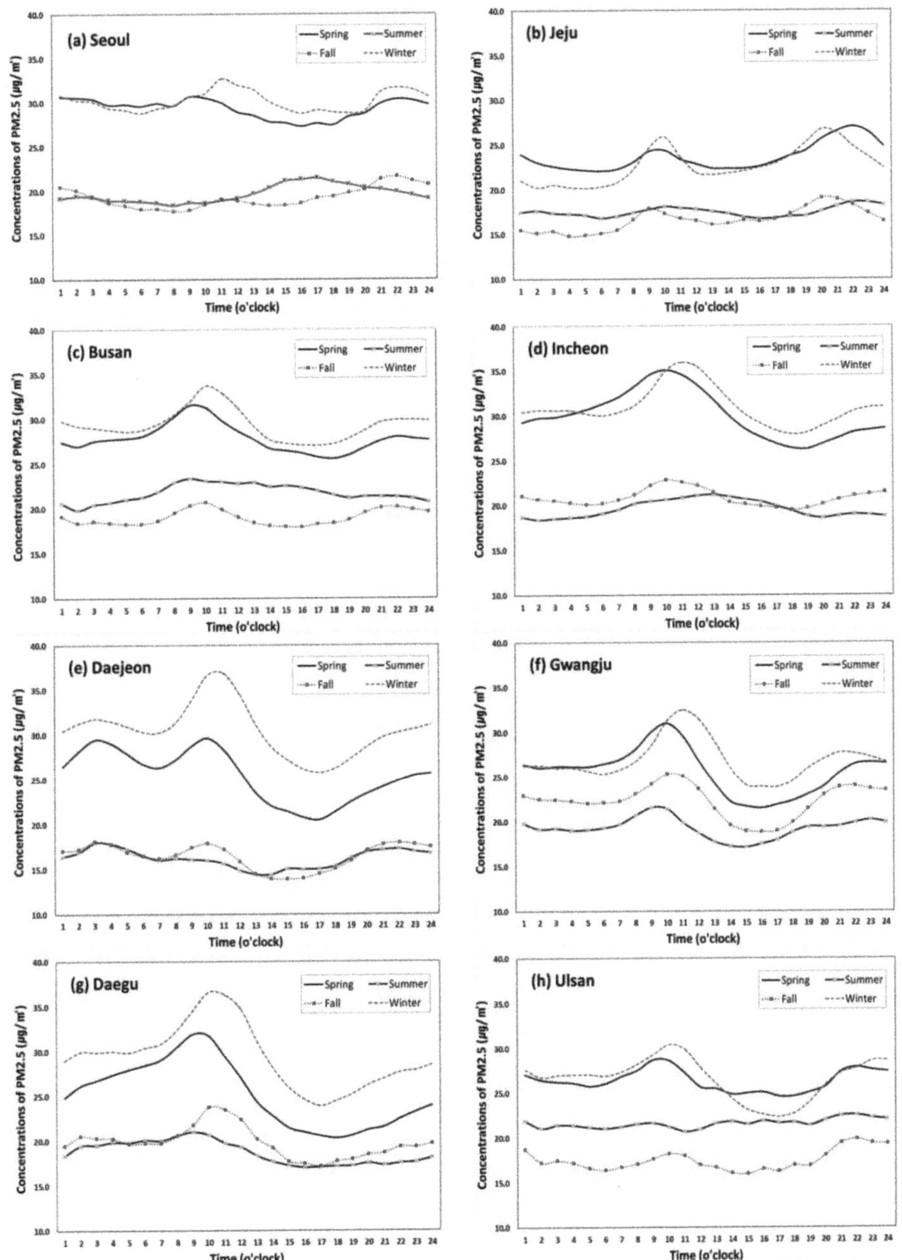

Figure 5. Diurnal variation of PM2.5 concentrations (μg/m^3) by season in each region (2015–2019).

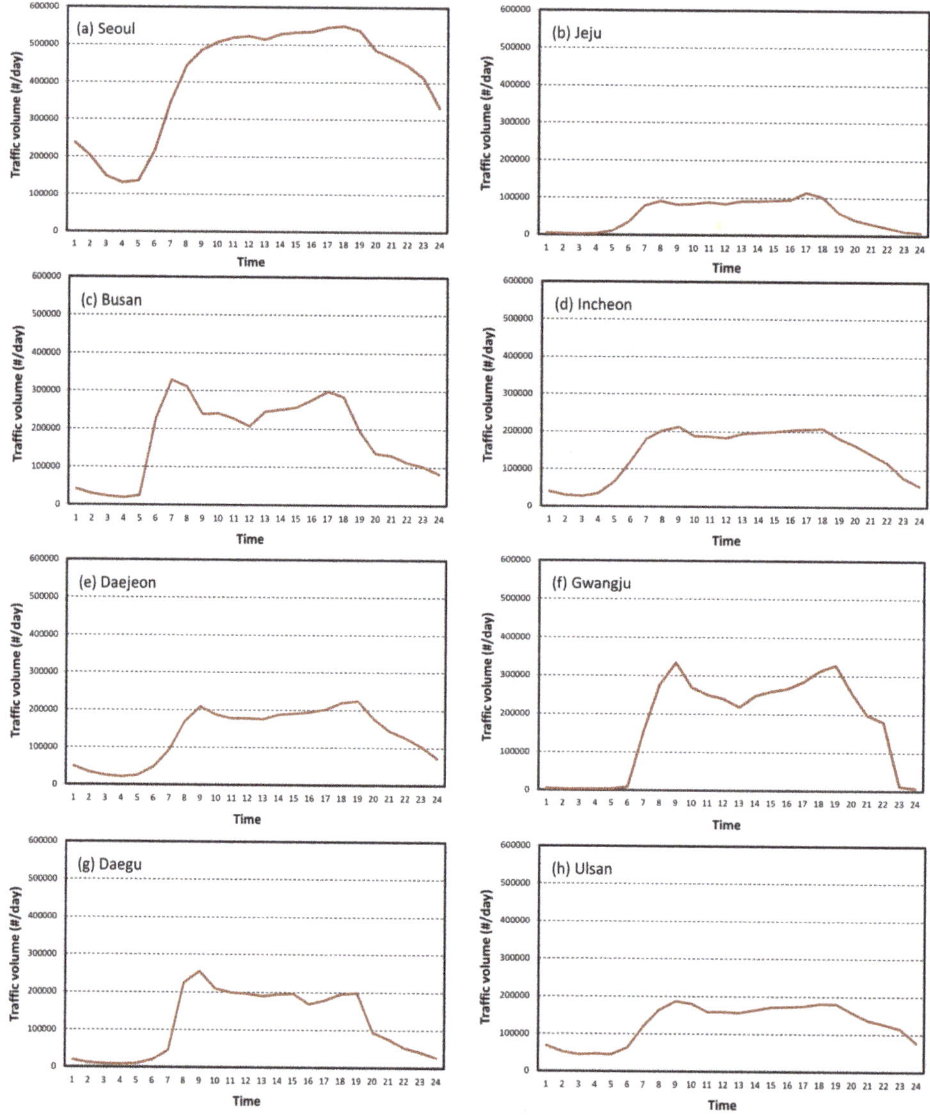

Figure 6. Diurnal variation of traffic volume in each region during 2016 and 2018.

4. Discussion

This study examined hourly and seasonal patterns of PM concentrations for multiple regions in South Korea based on the recent five-year data in order to find desirable and undesirable times for natural ventilation due to high levels of outdoor concentrations. Overall, the average concentrations of both PM10 and PM2.5 peaked in the morning, declined in the afternoon, and rebounded in the evening (particularly for PM2.5). The highest concentrations were generally found in spring (March to May) and winter (December to February), but PM2.5 concentrations were more concerning since they were above the KME's annual average standard of 15 $\mu g/m^3$ during peak times in all four seasons. Thus, it is generally advisable that natural ventilation is recommended during the afternoon (2–6pm) but should be avoided in the morning (9am to noon) or in the late evening (8–11pm). However, substantial differences were observed across the eight regions across the four seasons, suggesting intervention strategies tailored to regional climate and emission patterns. Moreover, the actual decision on whether ventilation is needed at a specific time should be based on a more comprehensive consideration of both indoor and outdoor air quality conditions.

Our results were consistent with previous studies in other countries showing a bimodal pattern with peaks during morning and evening rush hours [42]. This pattern was prominent in many other metropolitan cities such as New York, Los Angeles, Beijing, and London, implying that it was mostly attributed to anthropogenic activity, in particular motor vehicle traffic patterns [43–45]. Likewise, the morning peak noticed in this study could be mostly due to enhanced anthropogenic activity during commuting hours, while the afternoon valley is mainly due to a higher atmospheric mixing layer, which is beneficial for air pollution diffusion [46]. The morning peak was relatively more rigorous than the evening risk because, in South Korea, work starting time is relatively standard (around 8 or 8:30 am) but finishing time is more widely distributed due to variability in work schedule. This could partly explain why the PM levels in the evening were somewhat lower than those in the morning in this study.

The seasonal pattern of PM concentrations in South Korea is partly due to its unique geographic and climatic characteristics. The average PM10 concentration was highest in spring mainly because of yellow dust transported from northern China and the deserts of the Mongolian plateau by the prevailing westerlies [47]. The low PM levels during summer time are attributed to rainout and washout processes due to the rainy period as well as frequent typhoons, and rapid air circulation in the fall helps in reducing PM concentrations [24,48]. Generally, PM2.5 can stay longer in the air compared to PM10 due to a smaller size [49]. Local meteorological conditions (very low temperature, low wind speeds, surface layer inversions) and weak wind circulation during winter make it difficult to remove PM2.5 [50]. An increased usage of heating fuels during the winter raises PM levels, and they stay in the air for a longer period of time due to the cold surface of the earth and a low mixture rate [51,52]. This could explain why PM2.5 concentrations, unlike PM10, were higher in winter than spring since finer particles tend to be generated widely by man-made sources of emission such as solid fuel heating in winter [38].

In general, the level of PMs within a big metropolitan city is affected by traffic volume, point source pollution around the city (e.g., power plant), and weather conditions (especially wind direction) by season [53]. Increased fuel consumption for heating during winter could contribute to increasing the level of PM2.5, but other factors may also apply. The high winter concentrations of PM2.5 in Daejeon and Daegu (shown in Figure 4) are found to be associated with region-specific seasonal factors, such as the seasonal rise of bio-incineration via fuel use for household heating and cooking and emissions from petroleum-related industries [54,55] and traffic emission and gas-form pollutants [34]. The studies showed that the high level of PM2.5 during winter season in Daejeon area could be attributed to emission pollutants from two big coal-fired power plants located in the northwest of the city under the influence of the main wind direction during the winter season [56]. Similar patterns of PM2.5 concentrations over winter in Daegu could also be due to its geographic and climatic conditions. During winter, wind blowing from the northwest causes pollutants from the industrial complex located in the northwestern side of Daegu to enter the downtown area, and the particles remain stagnant in the area due to the lack of air circulation in the basin-shaped city [57]. Of course, a number of other geographic, climatic and socioeconomic factors specific to each region may contribute to PM trends as well [23].

5. Conclusions

Controlling air pollution is a daunting task in all industrialized countries because so many factors are involved and some of those even cross a country's border. Eliminating or reducing the sources of emission could bring other types of social conflict or dilemma which would require a long-term effort and investment to be resolved. Natural ventilation could be considered as a simple and low-cost solution to diminish the level of exposure to indoor air pollutants, but only when the outdoor air quality is acceptable. This study emphasizes the importance of thorough monitoring of ambient air quality to be used in providing guidelines for when indoor environments should be ventilated in different region. Despite substantial distinctions across the regions by season, this study provides some general suggestions that the time between 2–6pm is most suitable for natural ventilation but it is not desirable either in the morning or late evening, particularly during spring and winter. It also highlights PM2.5 concentrations surpassing the KME's annual average standard of 15 $\mu g/m^3$ during peak times in all four seasons and rebounding in size since 2019. Although further studies would be needed to confirm this suggestion and implement solutions in practice, the results of this study can be used as basic information for designing a comprehensive environmental health policy in consideration of dynamic exposure to air pollution.

Of course, it would be ideal to consider simultaneously both indoor and outdoor PM values in order to confirm the suggestions of this study. However, we believe that this study is valid in itself because indoor PM data are not readily available in most of the households and, even if available, the patterns would vary greatly vary according to building structure and lifestyle. This study aims to provide a broad guideline on appropriate times for natural ventilation based on outdoor air quality, particularly where indoor PM levels are unknown, but specific implementation should be carried out by considering indoor-outdoor dynamics pertaining to each building or household. Future study should be directed towards the confirmation of the trends found in this study based on actual experimental data on indoor and outdoor PM concentrations measured in sample buildings in multiple regions, instead of the public aggregated data. In addition, gathering more data on covariates indicating geographic, sociodemographic, climatic, industrial and traffic patterns in each region would enable multivariate analysis and modeling for the determinants of spatiotemporal and seasonal changes in PM concentrations. In spite of limitations and the inability to explain some patterns, this study can inform the public regarding the desirable or undesirable times for natural ventilation as well as outdoor activities in each season and region. Moreover, it can encourage policymakers to design season-specific environmental management strategies and practices tailored to regional climate and emission factors such as regional traffic monitoring networks and air quality alert systems for ventilation, which may promote a cost saving, by specialization of intervention, and enhance policy outcomes.

Author Contributions: Conceptualization, D.K. and S.S.; Data curation, H.-E.C.; Formal analysis, H.-E.C. and S.S.; Funding acquisition, S.S.; Methodology, D.K. and S.S.; Project administration, S.S. and W.-M.G.; Supervision, S.S. and W.-M.G.; Visualization, H.-E.C.; Writing—original draft, H.-E.C. and S.S.; Writing—review & editing, D.K. All authors read and approved the final manuscript.

Funding: This work was supported by Korea Environment Industry & Technology Institute (KEITI) through the Environmental Health Action Program, funded by Korea Ministry of Environment (MOE) (grant number: 2018001350005). This support is greatly appreciated.

Conflicts of Interest: The authors declare no conflict of interest.

References

1. Cincinelli, A.; Martellini, T. Indoor Air Quality and Health. *Int. J. Environ. Res. Public Health* **2017**, *14*, 1286. [CrossRef] [PubMed]
2. GBD 2017 Risk Factor Collaborators. Global, regional, and national comparative risk assessment of 84 behavioural, environmental and occupational, and metabolic risks or clusters of risks for 195 countries and territories, 1990–2017: A systematic analysis for the Global Burden of Disease Study 2017. *Lancet* **2018**, *392*, 1923–1994.
3. Kim, D.; Seo, S.; Min, S.; Simoni, Z.; Kim, S.; Kim, M. A Closer Look at the Bivariate Association between Ambient Air Pollution and Allergic Diseases: The Role of Spatial Analysis. *Int. J. Environ. Res. Public Health* **2018**, *15*, 1625. [CrossRef] [PubMed]
4. Baldacci, S.; Maio, S.; Cerrai, S.; Sarno, G.; Baïz, N.; Simoni, M.; Annesi-Maesano, I.; Viegi, G.; Study, H. Allergy and asthma: Effects of the exposure to particulate matter and biological allergens. *Respir. Med.* **2015**, *109*, 1089–1104. [CrossRef] [PubMed]

5. Lee, B.; Kim, B.; Lee, K. Air Pollution Exposure and Cardiovascular Disease. *Toxicol. Res.* **2014**, *30*, 71–75. [CrossRef]

6. Hernandez, A.M.; de Porras, D.G.R.; Marko, D.; Whitworth, K.W. The association between PM2. 5 and ozone and the prevalence of diabetes mellitus in the United States, 2002 to 2008. *J. Occup. Environ. Med.* **2018**, *60*, 594–602. [CrossRef]

7. Loomis, D.; Grosse, Y.; Lauby-Secretan, B.; El Ghissassi, F.; Bouvard, V.; Benbrahim-Tallaa, L.; Guha, N.; Baan, R.; Mattock, H.; Straif, K. International Agency for Research on Cancer Monograph Working Group IARC, The carcinogenicity of outdoor air pollution. *Lancet Oncol.* **2013**, *14*, 1262–1263. [CrossRef]

8. Kotzias, D. Indoor air and human exposure assessment–needs and approaches. *Exp. Toxicol. Pathol.* **2005**, *57*, 5–7. [CrossRef]

9. Li, Z.; Wen, Q.; Zhang, R. Sources, health effects and control strategies of indoor fine particulate matter (PM2.5): A review. *Sci. Total Environ.* **2017**, *586*, 610–622. [CrossRef]

10. Cho, K.; Kwon, S.; Sung, M.; Kim, S. Analysis of Indoor Air Quality in vulnerable facilities according to building characteristics. *J. Korea Inst. Healthc. Archit.* **2017**, *23*, 19–26.

11. Brasche, S.; Bischof, W. Daily time spent indoors in German homes – Baseline data for the assessment of indoor exposure of German occupants. *Int. J. Hyg. Environ. Health* **2005**, *208*, 247–253. [CrossRef] [PubMed]

12. Hoskins, J. Health Effects due to Indoor Air Pollution. *Indoor Built Environ.* **2003**, *12*, 427–433. [CrossRef]

13. Valavanidis, A.; Fiotakis, K.; Vlachogianni, T. Airborne particulate matter and human health: Toxicological assessment and importance of size and composition of particles for oxidative damage and carcinogenic mechanisms. *J. Environ. Sci. Health Part C* **2008**, *26*, 339–362. [CrossRef] [PubMed]

14. Stilianakis, N. Vulnerability to air pollution health effects. *Int. J. Hyg. Environ. Health* **2008**, *211*, 326–336.

15. Arvanitis, A.; Kotzias, D.; Kephalopoulos, S.; Career, P.; Cavallo, D.; Cesaroni, G.; De Brouwere, K. The INDEX-PM project: Health risks from exposure to indoor particulate matter. *Fresenius Environ. Bull.* **2010**, *19*, 2458–2471.

16. Maroni, M.; Seifert, B.; Lindvall, T. *Indoor Air Quality: A Comprehensive Reference Book*; Elsevier: Amsterdam, The Netherlands, 1995.

17. Chen, Y.; Tong, Z.; Malkawi, A. Investigating natural ventilation potentials across the globe: Regional and climatic variations. *Build. Environ.* **2017**, *122*, 386–396. [CrossRef]

18. Wargocki, P.; Sundell, J.; Bischof, W.; Brundrett, G.; Fanger, P.O.; Gyntelberg, F.; Hanssen, S.; Harrison, P.; Pickering, A.; Seppänen, O. Ventilation and health in nonindustrial indoor environments, Report from a European Multidisciplinary Scientific Consensus Meeting, Epidemiology. In Proceedings of the 7th REHVA World Congress and Clima 2000 Naples 2001 Conference, Milan, Italy, 15–18 September 2001; p. 94.

19. Chu, S.; Song, J. Indoor air quality for allergy prevention. *Pediatric Allergy Respir. Dis.* **2010**, *20*, 93–99.

20. Escombe, A.; Oeser, C.; Gilman, R.; Navincopa, M.; Ticona, E.; Pan, W.; Martinez, C.; Chacaltana, J.; Rodriguez, R.; Moore, D.; et al. Natural Ventilation for the Prevention of Airborne Contagion. *PLoS Med.* **2007**, *4*, e68. [CrossRef]

21. Ben-David, T.; Waring, M. Impact of natural versus mechanical ventilation on simulated indoor air quality and energy consumption in offices in fourteen U.S. cities. *Build. Environ.* **2016**, *104*, 320–336. [CrossRef]

22. Perino, M.; Heiselberg, P. Short-term airing by natural ventilation-modeling and control strategies. *Indoor Air* **2009**, *19*, 357–380. [CrossRef]

23. Lee, C.-T.; Chuang, M.-T.; Chan, C.-C.; Cheng, T.-J.; Huang, S.-L. Aerosol characteristics from the Taiwan aerosol supersite in the Asian yellow-dust periods of 2002. *Atmos. Environ.* **2006**, *40*, 3409–3418. [CrossRef]

24. Lee, H.-S.; Kang, B.-W. Influence on the Indoor Air Quality by Ambient Air during the Summer Season. *Korean Soc. Living Environ. Syst.* **1997**, *6*, 637–643.

25. Emmerich, S.; Dols, W.; Axley, J. *Natural Ventilation Review and Plan. for Design and Analysis Tools*; National Institute of Standards and Technology, U.S. Department of Commerce: Gaithersburg, MD, USA, 2001.

26. Branis, M.; Safranek, J.; Hytychova, A. Indoor and outdoor sources of size-resolved mass concentration of particulate matter in a school gym—Implications for exposure of exercising children. *Environ. Sci. Pollut. Res.* **2011**, *18*, 598–609. [CrossRef] [PubMed]

27. Kang, D.; Choi, D. A Preliminary Study to Evaluate the Impact of Outdoor Dust on Indoor Air by Measuring Indoor/Outdoor Particle Concentration in a Residential Housing Unit. *J. KIAEBS* **2015**, *9*, 462–469.

28. Dutton, S.; Banks, D.; Brunswick, S.; Fisk, W. Health and economic implications of natural ventilation in California offices. *Build. Environ.* **2013**, *67*, 34–45. [CrossRef]

29. Leung, D.Y.C. Outdoor-indoor air pollution in urban environment: Challenges and opportunity. *Front. Environ. Sci.* **2015**, *2*. [CrossRef]

30. Lee, D.; Choi, J.; Myoung, J.; Kim, O.; Park, J.; Shin, H.; Ban, S.; Park, H.; Nam, K. Analysis of a Severe PM2.5 Episode in the Seoul Metropolitan Area in South Korea from 27 February to 7 March 2019: Focused on Estimation of Domestic and Foreign Contribution. *Atmosphere* **2019**, *10*, 756. [CrossRef]

31. Park, S.; Kim, S.; Yu, H.; Lim, C.; Park, E.; Kim, J.; Lee, W. Developing an Adaptive Pathway to Mitigate Air Pollution Risk for Vulnerable Groups in South Korea. *Sustainability* **2020**, *12*, 1790. [CrossRef]

32. An, J.-H.; Oh, Y.-J.; Im, J.-Y.; Ahn, M.-S.; Hong, E.-J.; Son, B.-S. Concentration and risk assessment of indoor air quality in day care centers and postnatal care centers. *Korean Soc. Odor Indoor Environ.* **2018**, *17*, 337–345. [CrossRef]

33. Korea Ministry of Environment 7 Tips for High Concentration Fine Dust. 2017. Available online: http://www.me.go.kr/ (accessed on 5 June 2020).

34. Do, H.-S.; Choi, S.-J.; Park, M.-S.; Lim, J.-K.; Kwon, J.-D.; Kim, E.-K.; Song, H.-B. Distribution characteristics of the concentration of ambient PM-10 and PM-2.5 in Daegu area. *J. Korean Soc. Environ. Eng.* **2014**, *36*, 20–28. [CrossRef]

35. Yim, S.H.; Gu, Y.; Shapiro, M.A.; Stephens, B. Air quality and acid deposition impacts of local emissions and transboundary air pollution in Japan and South Korea. *Atmos. Chem. Phys.* **2019**, *19*, 13309–13323. [CrossRef]

36. Lee, H.; Lee, Y.J.; Park, S.Y.; Kim, Y.W.; Lee, Y. The improvement of ventilation behaviours in kitchens of residential buildings. *Indoor Built Environ.* **2012**, *21*, 48–61. [CrossRef]

37. Oh, H.J.; Nam, I.S.; Yun, H.; Kim, J.; Yang, J.; Sohn, J.R. Characterization of indoor air quality and efficiency of air purifier in childcare centers, Korea. *Build. Environ.* **2014**, *82*, 203–214. [CrossRef]

38. Heo, J.; Oh, S. Characterization of Annual PM2.5 and PM10 Concentrations by Real-time Measurements in Cheonan, Chungnam. *Korea Acad. Ind. Coop. Soc.* **2012**, *13*, 445–450. [CrossRef]

39. Jeong, J.-C.; Lee, P.S.-H. Spatial distribution of particulate matters in comparison with land-use and traffic volume in Seoul, Republic of Korea. *J. Cadastre Land Inf.* **2018**, *48*, 123–138.

40. Oh, I.; Bang, J.-H.; Kim, S.; Kim, E.; Hwang, M.-K.; Kim, Y. Spatial distribution of air pollution in the Ulsan metropolitan region. *J. Korean Soc. Atmos. Environ.* **2016**, *32*, 394–407. [CrossRef]

41. Kim, E.; Bae, C.; Yoo, C.; Kim, B.-U.; Kim, H.C.; Kim, S. Evaluation of the Effectiveness of Emission Control Measures to Improve PM 2.5 Concentration in South Korea. *J. Korean Soc. Atmos. Environ.* **2018**, *34*, 469–485. [CrossRef]

42. Adaes, J.; Pires, J. Analysis and Modelling of PM2. 5 Temporal and Spatial Behaviors in European Cities. *Sustainability* **2019**, *11*, 6019. [CrossRef]

43. DeGaetano, A.; Doherty, O.M. Temporal, spatial and meteorological variations in hourly PM2.5 concentration extremes in New York City. *Atmos. Environ.* **2004**, *38*, 1547–1558. [CrossRef]

44. Moore, K.; Verma, V.; Minguillon, M.; Sioutas, C. Inter- and Intra-Community Variability in Continuous Coarse Particulate Matter (PM10–2.5) Concentrations in the Los Angeles Area. *Aerosol Sci. Technol.* **2010**, *44*, 526–540. [CrossRef]

45. Liu, Z.; Hu, B.; Wang, L.; Wu, F.; Gao, W.; Wang, Y. Seasonal and diurnal variation in particulate matter (PM 10 and PM 2.5) at an urban site of Beijing: Analyses from a 9-year study. *Environ. Sci. Pollut. Res.* **2015**, *22*, 627–642. [CrossRef] [PubMed]

46. Wang, Y.; Zhang, X.; Sun, J.; Zhang, X.; Che, H.; Li, Y. Spatial and temporal variations of the concentrations of PM10, PM2.5 and PM1 in China. *Atmos. Chem. Phys.* **2015**, *15*, 13585–13598. [CrossRef]

47. Kim, H. Analysis of environmental policy in US and KOREA. *Reform. Environ. Legis.* **2002**, *22*, 401–402.

48. Kim, B.; Kim, D. Studies on the environmental behaviors of ambient PM2. 5 and PM10 in Suwon Area. *J. Korean Soc. Atmos. Environ.* **2000**, *16*, 89–101.

49. Baron, P.A.; Willeke, K. *Aerosol Measurement: Principles, Techniques, and Applications*, 3rd ed.; Wiley: Hoboken, NJ, USA, 2001.

50. Przybysz, A.; Nersisyan, G.; Gawroński, S.W. Removal of particulate matter and trace elements from ambient air by urban greenery in the winter season. *Environ. Sci. Pollut. Res.* **2019**, *26*, 473–482. [CrossRef]

51. Korea Ministry of Environment the Government, Confirmation and Announcement of Special Measures for Fine Dust Management. Available online: http://www.me.go.kr/ (accessed on 5 June 2020).

52. Kim, S.; Kim, C. The physico-chemical character of aerosol particle in Seoul metropolitan area. *Seoul Inst.* **2008**, *9*, 23–33.

53. Kassomenos, P.A.; Vardoulakis, S.; Chaloulakou, A.; Paschalidou, A.K.; Grivas, G.; Lumbreras, J. Study of PM10 and PM2. 5 levels in three European cities: Analysis of intra and inter urban variations. *Atmos. Environ.* **2014**, *87*, 153–163. [CrossRef]
54. Kim, H.; Jung, J.; Lee, J.; Lee, S. Seasonal Characteristics of Organic Carbon and Elemental Carbon in PM2.5 in Daejeon. *J. Korean Soc. Atmos. Environ.* **2015**, *31*, 28–40. [CrossRef]
55. Sung, M.-Y.; Park, J.-S.; Kim, H.-J.; Jeon, H.-E.; Hong, Y.-D.; Hong, J.-H. The Characteristics of Element Components in PM2.5 in Seoul and Daejeon. *Korean Soc. Environ. Anal.* **2015**, *18*, 49–58.
56. Ju, J.; Youn, D. Statistically Analyzed Effects of Coal-Fired Power Plants in West Coast on the Surface Air Pollutants over Seoul Metropolitan Area. *J. Korean Earth Sci. Soc.* **2019**, *40*, 549–560. [CrossRef]
57. Ju, J.; Hwang, I. A study for spatial distribution of principal pollutants in Daegu area using air pollution monitoring network data. *J. Korean Soc. Atmos. Environ.* **2011**, *27*, 545–557. [CrossRef]

International Journal of
*Environmental Research
and Public Health*

Article

Evolution of Urban Haze in Greater Bangkok and Association with Local Meteorological and Synoptic Characteristics during Two Recent Haze Episodes

Nishit Aman [1,2], Kasemsan Manomaiphiboon [1,2,*], Natchanok Pala-En [3], Eakkachai Kokkaew [1,2], Tassana Boonyoo [4], Suchart Pattaramunikul [4], Bikash Devkota [1,2] and Chakrit Chotomosak [5]

1   The Joint Graduate School of Energy and Environment, King Mongkut's University of Technology Thonburi, Bangkok 10140, Thailand; aman.nishit@gmail.com (N.A.); eakkachai.ko@gmail.com (E.K.); devkota.bikky@gmail.com (B.D.)
2   Center of Excellence on Energy Technology and Environment, Ministry of Education, Bangkok 10140, Thailand
3   Pollution Control Department, Ministry of Natural Resources and Environment, Bangkok 10400, Thailand; natpalaen@gmail.com
4   Traffic and Transport Development and Research Center, King Mongkut's University of Technology Thonburi, Bangkok 10140, Thailand; tassana.boo@gmail.com (T.B.); suchart.pat@gmail.com (S.P.)
5   Department of Geography, Chiang Mai University, Chiang Mai 50200, Thailand; chotamonsak@gmail.com
*   Correspondence: kasemsan_m@jgsee.kmutt.ac.th or kasemsan.jgsee@gmail.com; Tel.: +66-2-470-7331

Received: 18 November 2020; Accepted: 15 December 2020; Published: 18 December 2020

Abstract: This present work investigates several local and synoptic meteorological aspects associated with two wintertime haze episodes in Greater Bangkok using observational data, covering synoptic patterns evolution, day-to-day and diurnal variation, dynamic stability, temperature inversion, and back-trajectories. The episodes include an elevated haze event of 16 days (14–29 January 2015) for the first episode and 8 days (19–26 December 2017) for the second episode, together with some days before and after the haze event. Daily $PM_{2.5}$ was found to be 50 μg m^{-3} or higher over most of the days during both haze events. These haze events commonly have cold surges as the background synoptic feature to initiate or trigger haze evolution. A cold surge reached the study area before the start of each haze event, causing temperature and relative humidity to drop abruptly initially but then gradually increased as the cold surge weakened or dissipated. Wind speed was relatively high when the cold surge was active. Global radiation was generally modulated by cloud cover, which turns relatively high during each haze event because cold surge induces less cloud. Daytime dynamic stability was generally unstable along the course of each haze event, except being stable at the ending of the second haze event due to a tropical depression. In each haze event, low-level temperature inversion existed, with multiple layers seen in the beginning, effectively suppressing atmospheric dilution. Large-scale subsidence inversion aloft was also persistently present. In both episodes, $PM_{2.5}$ showed stronger diurnality during the time of elevated haze, as compared to the pre- and post-haze periods. During the first episode, an apparent contrast of $PM_{2.5}$ diurnality was seen between the first and second parts of the haze event with relatively low afternoon $PM_{2.5}$ over its first part, but relatively high afternoon $PM_{2.5}$ over its second part, possibly due to the role of secondary aerosols. $PM_{2.5}/PM_{10}$ ratio was relatively lower in the first episode because of more impact of biomass burning, which was in general agreement with back-trajectories and active fire hotspots. The second haze event, with little biomass burning in the region, was likely to be caused mainly by local anthropogenic emissions. These findings suggest a need for haze-related policymaking with an integrated approach that accounts for all important emission sectors for both particulate and gaseous precursors of secondary aerosols. Given that cold surges induce an abrupt change in local meteorology, the time window to apply control measures for haze is limited, emphasizing the need for readiness in mitigation responses and early public warning.

Keywords: urban haze; temperature inversion; Obukhov length; HYSPLIT; biomass burning; cold surge, emission

---

1. Introduction

Particulate matter (PM) with a size less than or equal to 2.5 µm ($PM_{2.5}$) is an environmental concern worldwide. Suspension of such particulate matter in the atmosphere (known as aerosols) affects human health, atmospheric visibility and also impacts weather and climate both directly and indirectly [1]. The aerosols are either emitted directly in the atmosphere (known as primary sources) from combustion, wind-borne dust, sea spray, volcanic emission, and biogenic aerosol or formed in the atmosphere by conversion of primary precursor gases to secondary particles through nucleation and complex multiphase chemical reactions. High PM pollution in any place always depends upon the complex interplay between local emissions, secondary particle formation, along with local and synoptic meteorology [1,2]. Around 58% of the world population lives in areas with $PM_{2.5} > 35$ µg m^{-3} (in terms of daily average according to the WHO Interim Target 1) [3], many of which are highly urbanized areas or large cities. Although PM pollution episodes (sometimes known as haze episodes) are limited to few days to weeks, exposure to high PM levels even for shorter times can have an effect on human health and ecosystems [3].

Haze pollution has been studied globally to understand its formation and evolution mechanism [4–7], potentials sources contribution [5,8,9], mitigation strategies [10,11], and early warning or forecasting [12]. A haze episode may be induced by one or a number of factors combined, which encompasses emissions, secondary aerosols [13–17], and atmospheric transport [8,17], with unfavorable weather conditions acting as an accelerating factor [18–25]. The effects of the atmospheric boundary layer (ABL) structure, near-surface atmospheric stability, and synoptic conditions have received the attention given that they strongly dictate how haze and its associated thermal and dynamical processes evolve with time in the lower part of the troposphere [23–25]. Quan et al. (2013) [19] and Petaja et al. (2016) [20] suggested a positive feedback cycle for heavy air pollution where heat flux decrease significantly due to decreased solar radiation blocked by the haze layer, which in turn further decreases ABL height and trap air pollutant within. Tie et al. (2017) [21], and Liu et al. (2018) [22] reported that decreased ABL height leads to increased relative humidity, which enhances secondary aerosol formation and also promotes the hygroscopic growth of aerosols and light scattering. The latter effect reduces incident solar radiation more. Temperature inversion at different levels of the troposphere may be caused by radiation, advection, large-scale subsidence, etc. Low-level temperature inversion layers effectively reduce the atmospheric volume available for diluting air pollutants while weak winds poorly ventilate them out of a polluted area [23–25]. Turbulence is another important factor as the capability of mixing airborne constituents vertically in the ABL. Low turbulence, typically under stable conditions, suppresses vertical mixing, which allows pollutants and/or precursors to accumulate at near-surface levels [23,26]. Synoptic weather is another relevant aspect because it sets favorable or unfavorable background conditions to haze formation and evolution and influences local or urban-scale weather processes as well [23–27].

Bangkok, the capital of Thailand, and its five neighboring provinces (Samut Prakan, Samut Sakhon, Nonthaburi, Nakhon Pathom, and Pathum Thani) are collectively known as Greater Bangkok (GBK) or Bangkok Metropolitan. It is one of the largest urban agglomerations in Southeast Asia. It has experienced haze pollution typically found in the dry season, posing great concern to the general public and challenges to the local and central governments for mitigation and prevention. $PM_{2.5}$ exceeds the daily (i.e., 24-h average) national ambient air quality standard (NAAQS) of 50 µg m^{-3} several times per year, according to the Pollution Control Department (PCD) [28], particularly in the dry season. Several PM-related studies in GBK have been conducted e.g., [29–38], see Table S1 in Supplementary Materials for details, ranging from source apportionment, chemical characterization, emission inventory, and human health. On-road vehicles (i.e., traffics) and biomass burning were identified as the major $PM_{2.5}$ sources [29–32]. In the dry season, agricultural burning to clear crop residues on lands within GBK and its vicinity and forest fires in the northern region are generally intensified, and smoke can also be dispersed or transported to GBK [31–34]. However, biomass burning contributes little to air pollution in GBK during the summer due to a shift in the prevailing winds [35]. Pham et al. (2008) [36] estimated gaseous and particulate emissions from industrial and power plant sectors, finding the central and central regions (among all regions in Thailand) to have the largest intensities and shares. Secondary aerosols were also reported as a nonnegligible contributor [30,32]. Effects of increased $PM_{2.5}$ on health have been emphasized and quantified [37,38]. Although these past studies provide useful information of the PM sources and PM effects for the study area in question, elevated haze and its associated meteorological dependence are still lacking or little addressed. Accordingly, this study aims to fill this knowledge gap by investigating how urban haze is influenced by meteorology at both local and synoptic scales. Specifically, we seek to understand how events of elevated haze (in terms of $PM_{2.5}$) temporally evolve with meteorological factors using several observational datasets combined, with particular attention to cold surge as a synoptic disturbance relevant to the region and here suspected to induce favorable conditions for haze to elevate. Here, an intensive observational analysis was performed and discussed for two recent wintertime haze episodes in GBK using data from various sources to examine the association of haze with local and synoptic weather conditions.

2. Data and Methods

2.1. Study Area

Greater Bangkok (GBK) is located in the lower part of Central Thailand. It currently has a registered population of 11 million and an area of 7762 km^2 [39]. It is the largest national hub of the economy, accounting for 46% of its total gross domestic product (GDP) [40]. It has a complex mixed (built and natural) landscape with the co-existence of commercial, residential, agricultural, and industrial areas [41]. The overall terrain of GBK is generally flat with limited heights (<10 m above mean sea level or MSL). Its general climate is tropical and humid and governed mainly by the northeast monsoon (November–February as the winter) and the southwest monsoon (May–October as the wet season) [42]. The former monsoon brings cool, dry air from continental mid-latitudes over which persistent strong high-pressure systems are present. The latter monsoon brings moist air from the Indian Ocean and the Gulf of Thailand, causing abundant rain in most parts of Thailand. The monsoon trough or intertropical convergence zone (ITCZ), which moves along the north-south direction, and tropical cyclones developed in the North Indian and Western Pacific Ocean Basins can modulate rain at a sub-seasonal scale. Importantly, a cold surge is a synoptic phenomenon, characterized by a transient southward propagation or extension of a high-pressure system from mid-latitudes (i.e., mainland China and East Asia) to the Indochina Peninsula and the equatorial South China Seas [43–45]. During the winter, cold surges occur episodically more often, strengthening the northeast monsoon. The arrival of a cold surge typically brings strong winds due to high-pressure gradients with an abrupt drop in temperature. Once it weakens, the cold surge recedes back or dissipates. The transitional period between the two monsoons (March–April) has relatively warm conditions, corresponding to the summer season. Here, the winter and summer combined are called the dry season [42].

2.2. Data

The PCD is the main government agency that administers air quality monitoring stations across Thailand. Here, hourly $PM_{2.5}$ and PM_{10} data for the years 2015–2017 at three air quality stations (P27, P59, and P61) in the study area were requested and obtained (Table 1 and Figure 1c). P27 is located near a busy major highway (by about 70 m). P59 is in a semi-general area but not far from a busy local street and a major expressway to the west by 0.6 km. P61 is in a general area with no major road nearby within 1 km, representatively selected as the key station to support several parts of the analysis. It is noted that the PM$_{2.5}$ data at P59 are available only from April 2015. Given missing data and discontinuity in meteorological measurements at these three stations, the 100-m tower (M6) of the PCD at Techno Thani in Pathum Thani was the main source of meteorological data for use. The tower measures air temperature (T) at 2 m, 50 m, 75 m, and 100 m, wind speed (WS) and wind direction (WD) at 10 m, 50 m, and 100 m, and other near-surface variables, i.e., relative humidity (RH), rain (RN) and global radiation (GR). PM$_{2.5}$ and PM$_{10}$ are real-time monitored using the standard beta-ray method at 3 m above ground level (AGL). Data are intensively quality controlled/assured by the PCD internally before distribution. The detectable limits or probable ranges of data are PM$_{2.5}$ and PM$_{10}$ (3 to 1000 μg m^{-3}), temperature (−5 to 50 °C), relative humidity (0 to 100%), wind speed (0 to 50 ms^{-1}), wind direction (0° to 360°), rain (0 to 1000 mm h^{-1}) and global radiation (0 to 1000 W m^{-2}) [46,47]. We applied these ranges for data screening and also removed suspicious or erratic values if found visually. Relative humidity data at the M6 tower are absent throughout 2017. Simple linear-regression extrapolation using the data at P59 was employed to gap-fill them (details not shown). The tower is in a suburban/rural area, with most of the land near or surrounding the tower being water and paddy fields. Built-up areas are also well present over its northeastern quadrant within 2 km. Since many buildings taller than 10 m are also in their proximity of 100 m, 10-m wind data were discarded. Upper-air sounding data at the Bang Na weather station of the Thai Meteorological Department (TMD) was obtained (available at http://weather.uwyo.edu/upperair/sounding.html). However, radiosonde soundings are routinely operated only at 7 local time (LT) (i.e., 0 UTC), with pilot balloons alone at 1, 13, and 19 LT. The upper-air sounding and data of the TMD typically follow the standard quality assurance of the World Meteorological Organization (WMO). Here, radiosonde-based upper-air data up to a height of 4 km were considered and extracted. To convert hourly data to a daily scale for air quality and meteorological variables, 24-h (1–24 LT) averaging was typically applied. For global radiation, a period of 11–16 LT was used to represent late-morning to mid-afternoon hours. All computations and statistical tests were performed using software R (R development core team, 2019) [48]. In any statistical calculation, 50% of data as valid/ non-missing was necessarily required as the minimum threshold.

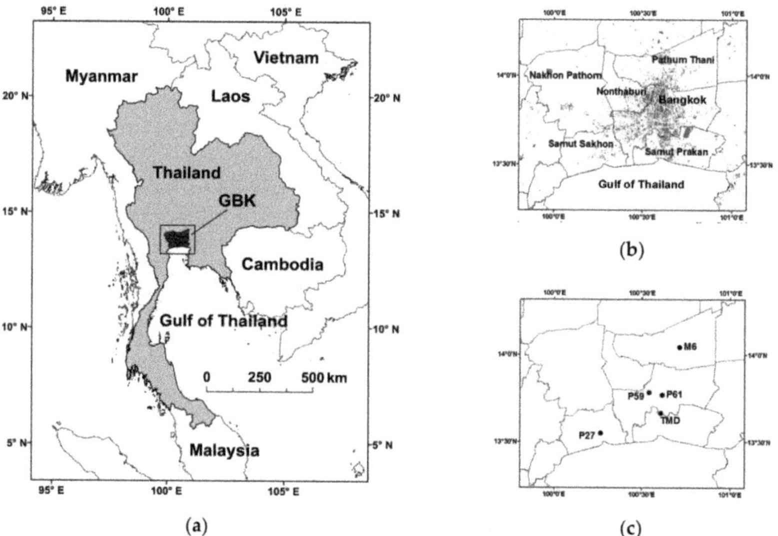

Figure 1. (**a**) Thailand; (**b**) Greater Bangkok and its provinces; (**c**) monitoring stations. In (**b**), the urban built-up areas (urban residential, industrial, and commercial classes combined) are shown in shaded

gray, based on LDD (Land Development Department) (2016) [41]. In (**c**) P27, P59, and P61 are the air quality stations at Samut Sakhon Witthayalai School, Public Relation Department, and Bodindecha School, respectively, M6 is the 100-m tower at Techno Thani, and the Thai Meteorological Department (TMD) is the standard weather station (WMO No. 48453) at Bang Na District.

Table 1. Monitoring stations considered in this study.

Station	Province	Variables	Station Type (Background)
P27	Samut Sakhon	Hourly $PM_{2.5}$, and PM_{10}	Surface air quality (roadside or semi-general)
P59	Bangkok	Hourly $PM_{2.5}$, and PM_{10}, RH	Surface air quality (semi-general)
P61	Bangkok	Hourly $PM_{2.5}$ and PM_{10}	Surface air quality (general)
M6	Pathum Thani	Hourly $T2$, $T50$, $WS50$, $WS100$, $WD50$, $WD100$, RH, RN, and GR	100-m tower (general)
TMD	Bangkok	T (at different heights)	Sounding at 7 LT

Note: $PM_{2.5}$ and PM_{10} ($\mu g\ m^{-3}$): particulate matter with size not larger than 2.5 μm and 10 μm, respectively; $T2$ and $T50$ (°C): temperature at 2 m and 50 m, respectively; $WS50$ and $WS100$ ($m\ s^{-1}$): wind speed at 50 m and 100 m, respectively; $WD50$ and $WD100$ (degrees from the north): wind direction at 50 m and 100 m, respectively; RH (%): relative humidity; RN (mm): rain; GR ($W\ m^{-2}$): global radiation; the stations have the terrain elevations of 2–4 m MSL.

2.3. Selection of Haze Episodes

A haze day is here defined as the day with $PM_{2.5}$ exceedance (i.e., daily $PM_{2.5}$ level $\geq 50\ \mu g\ m^{-3}$) registered at one station at least. Then, a haze event is simply defined as the period of consecutive haze days. As seen from Figure 2a (also see Figure S1 in Supplementary Materials), December–February is typically the period when haze intensifies. Given emphasis to extended (e.g., haze days over a week or longer) severe episodes and the amount of the valid surface and upper-air data, two haze episodes (EP1 and EP2) were representatively chosen (Figure 2b,c). In EP1, it comprises the following three periods in sequence: pre-HZ1, HZ1, and post-HZ1. The haze event (namely, HZ1) spans 16 days (14–29 January 2015). Pre-HZ1 and post-HZ1 are the periods before (7 days) and after (4 days) HZ1, respectively, when $PM_{2.5}$ was at relatively low levels. Similarly, EP2 covers pre-HZ2, HZ2, and post-HZ2. The haze event (namely, HZ2) spans 8 days (19–26 December 2017), with 6-day pre-HZ2 and 5-day post-HZ2.

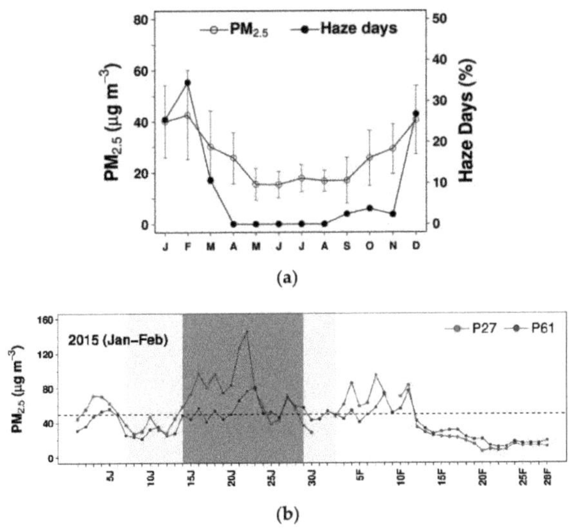

Figure 2. *Cont.*

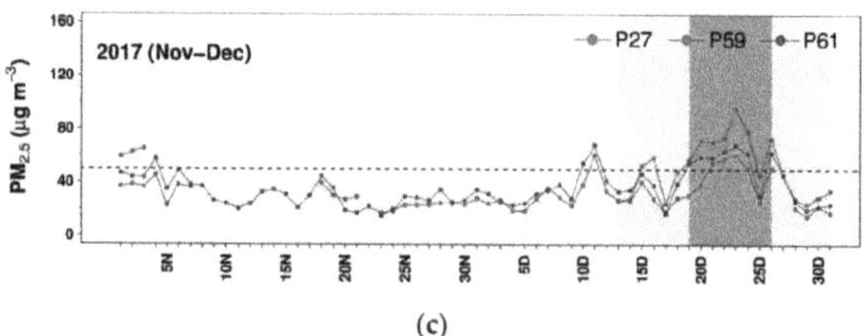

(c)

Figure 2. (**a**) Monthly particulate matter (PM) with a size less than or equal to 2.5 µm (PM$_{2.5}$) and number of haze days observed at P61 over 2015–2017; (**b**) daily PM$_{2.5}$ during haze episode 1 (EP1); (**c**) daily PM$_{2.5}$ during haze episode 2 (EP2). In (**a**), the filled and unfilled circles are the averages, the vertical bars are the standard deviations, and the *x*-axis labels (J, F, M, … , N, and D) denote the months of the year. In (**b,c**), the suffixes N, D, J, and F on the *x*-axis labels correspond to November–February, respectively. In (**b,c**), the dark-gray shading marks the HZ periods, and the light-gray shading marks the pre-haze event (HZ) and post-HZ periods for each haze episode.

2.4. Synoptic Patterns

The 6-hourly 0.5°-resolution Climate Forecast System Version 2 (CFSv2) reanalysis data [49] were used to provide sea-level pressure at 13 LT for all days in EP1 and EP2 to help construct daily synoptic surface charts. Moreover, satellite infrared images (at the same local time) captured by geostationary Himawari-8 (available at http://weather.is.kochi-u.ac.jp/sat/gms.sea/) were used to examine the presence of low and high clouds over the central region of Thailand, the ITCZ, and tropical cyclones. It is noted that the constructed CFSv2-based surface charts were also compared with the corresponding TMD surface charts (available at https://www.tmd.go.th/en/weather_map.php), both of which were found to be similar. A total of four simple synoptic patterns (numbered as 0, 1, 2, and 3) were proposed here to help support the analysis, using our visual examination of the constructed charts. The representative surface charts for EP1 and EP2 are shown in Figures S2 and S3 (Supplementary Materials), respectively. The four synoptic patterns classified are as follows:

a. Pattern 0: No distinct synoptic features over the Indochina. The ITCZ tends to stay over the central or lower portions of the Gulf of Thailand. On some days over GBK and the central region, clouds develop or scatter from the ITCZ edge;

b. Pattern 1: A cold surge propagates southward, and its front reaching the Indochina with relatively weak pressure gradients (i.e., weak winds) over the central region. For the ITCZ and clouds, same as Pattern 0;

c. Pattern 2: When the cold surge continues to propagate southward and its front reaching in the central region with moderate-to-strong pressure gradients (i.e., stronger winds). Clouds over GBK and central region are quite limited;

d. Pattern 3: When the cold surge weakens over the central region, or its front recedes northward or dissipates. For the ITCZ and clouds, same as Pattern 0.

2.5. Temperature Inversion and Obukhov Length

A temperature inversion refers to an atmospheric condition when the air temperature increases with height, and its presence can restrict the volumetric dilution of air pollutants [2,50]. Using the radiosonde-based upper-air data (available only at 7 LT), temperature inversion layers were identified. Those only found over the heights of 100–1500 m are of interest here. Those below 100 m were excluded since they are typically induced by continuous radiative surface cooling over the nighttime and early morning hours. The upper limit of 1500 m conservatively marks the ABL thickness. A single inversion layer is given as all successive vertical levels from the sounding with temperature monotonically increasing with height. Inversion intensity (*IV*) is a simple parameter used to indicate the extent of difficulty or blockage to which air pollutants penetrate an inversion layer, which was here computed as the rate of temperature change (°C per 100 m depth) from the bottom to the top of the inversion layer, similar to Dai et al. (2020) [7]. Only inversion layers with $IV > 0.1$ °C per 100 m were considered. Obukhov length (L) is an important measure of near-surface dynamic stability [2,18,23], suggesting the capability of vertical mixing for air pollutants between the surface and higher levels. Its positive/negative values correspond to stable/unstable conditions. The smaller magnitude of L, the larger degree of stability/instability. A very large L in magnitude corresponds to the neutral condition. The term "dynamic" implies that both mechanical and thermal turbulence production processes are taken into account. Following the Monin–Obukhov similarity theory [18,23], L is computed by

$$L = \frac{1}{2}\frac{(T_{v1} + T_{v2})u_*^2}{kg\theta_*}, \tag{1}$$

$$u_* = k\,U(z_u)\left[ln\!\left(\frac{z_u}{z_0}\right) - \Psi_M\!\left(\frac{z_u}{L}\right) + \Psi_M\!\left(\frac{z_0}{L}\right)\right]^{-1}, \text{ and} \tag{2}$$

$$\theta_* = k\,(\theta_{v2} - \theta_{v1})\left[ln\!\left(\frac{z_{\theta v2}}{z_{\theta v1}}\right) - \Psi_H\!\left(\frac{z_{\theta v2}}{L}\right) + \Psi_H\!\left(\frac{z_{\theta v1}}{L}\right)\right]^{-1}, \tag{3}$$

where u_* is the frictional velocity, θ_* is the temperature scale, k is the von Karman constant (0.4), g is the acceleration due to gravity, z_0 is the roughness length, z_u is the single measurement height of wind (here, 50 m or 100 m separately), $U(z_u)$ is the hourly wind speed, $z_{\theta v1}$ and $z_{\theta v2}$ are the two measurement heights of temperature (here, 2 m and 50 m, respectively), θ_{v1} and θ_{v2} are the virtual potential temperature at those two heights, T_{v1} and T_{v2} are the virtual temperatures at those two heights, respectively, and Ψ_M and Ψ_H are the Businger stability correction functions for wind and temperature, respectively. As in Kamma et al. (2020) [51], the surrounding area of the M6 tower within a 2 km radius was assessed using Google Earth (https://www.google.com/earth/) (Google, Mountain View, CA, USA) and visually examined, finding ponds and paddy fields being dominant with built-up areas present in its northeast quadrant. Using the classification by Stewart and Oke (2012) [52], the approximate local climate zone is "sparsely built" with terrain roughness (or Davenport) class 5, whose roughness length (z_0) equals 0.25. The concept of virtual (potential) temperature is necessary for humidity correction, which was implemented using the "aiRthermo" package in R [53]. In doing so, surface pressure data were required, extracted from 3-hourly 0.25°-resolution global land data assimilation system (GLDAS) [54] (available at https://giovanni.gsfc.nasa.gov/giovanni/), and then linearly interpolated to an hourly scale. If pressure at any higher levels was needed, the hydrostatic adjustment was applied. We attempted to compute hourly L with $z_u = 50$ m and $z_u = 100$ m separately and found a very high correlation (0.99) between the results from the two cases. Thus, the average L values over both cases were used.

2.6. Back-Trajectories

To investigate the potential transport of air pollutants from nearby and far areas [23,24], daily kinematic back-trajectories were simulated for all days in both episodes (EP1 and EP2) by the Hybrid Single-Particle Lagrangian Integrated Trajectory model (HYSPLIT) of the National Oceanic and Atmospheric Administration (NOAA) [55]. Each back-trajectory starts at 13 LT (as a typical midday time with developed ABL) and at 500 m AGL (as a typical mid-ABL height) and migrates backward in time for 48 h. HYSPLIT was run online (at https://www.ready.noaa.gov/HYSPLIT.php) using hourly 0.5°-resolution global data assimilation system (GDAS) data for driving wind fields. Given that biomass burning (agricultural burning and forest fires) in Upper Southeast Asia is well present in the dry season [32–34], daily 1-km active fire hotspots detected by the MODIS (Moderate Resolution Imaging Spectroradiometer) sensors onboard of both Terra and Aqua satellites (MCD14ML Collection 6) [56] (available at https://firms.modaps.eosdis.nasa.gov/download/) were downloaded for both episodes. The fire hotspots were then summed and gridded to 0.5° according to the pre-HZ, HZ, and post-HZ periods for each episode.

3. Results and Discussion

The urban haze in GBK is generally associated with multiple factors and their interdependence or interplays, ranging from emissions from local sources and biomass burning, mid- and long-range transport, secondary aerosols, and meteorological conditions at both local and synoptic scales. Here, the last factor is our main focus, for which general and distinct meteorological features and how they are coupled with the urban haze evolving during the selected two haze episodes are described.

3.1. Haze Episode EP1

This episode (EP1) occurred mostly in January 2015, and the haze event (HZ1) spans 14–29 January. Figure 3 displays the day-to-day variation of $PM_{2.5}$, $PM_{2.5}/PM_{10}$ ratio, temperature, relative humidity, wind speed, global radiation, rain, and synoptic pattern in the episode. Every variable was of 24-h average, except for global radiation (11–16 LT). For $PM_{2.5}/PM_{10}$, its daily values were of the 24-h average of the ratio of hourly $PM_{2.5}$ to hourly PM_{10}. As seen from the figure, $PM_{2.5}$ was relatively high in HZ1 but low during both pre-HZ1 and post-HZ1. It showed two peaks, 80.7 µg m^{-3} on 23 January and 68.3 µg m^{-3} on 27 January. $PM_{2.5}/PM_{10}$ did not appear to vary much (ranging between 0.45 and 0.66), being slightly higher in HZ1 than pre-HZ1 and post-HZ1, suggesting fine PM mode to increase when the haze was more developed. Temperature and relative humidity were 28.2 °C and 62.9%, at the start of EP1, respectively. Both decreased to 22.0 °C and 47.6%, respectively, at the start of HZ1, as caused by the synoptic change from Pattern 1 to Pattern 2 (i.e., cold surge reaching GBK with cool, dry air), but later climbed up continuously until HZ1 ended, corresponding synoptically to the weakening cold surge or its eventual dissipation. Winds appeared to follow the synoptic patterns, which were relatively strong during the cold surge arrival (i.e., Pattern 2), as seen on 11–14 January and became weaker on the other days, particularly during HZ1, supporting the buildup of haze.

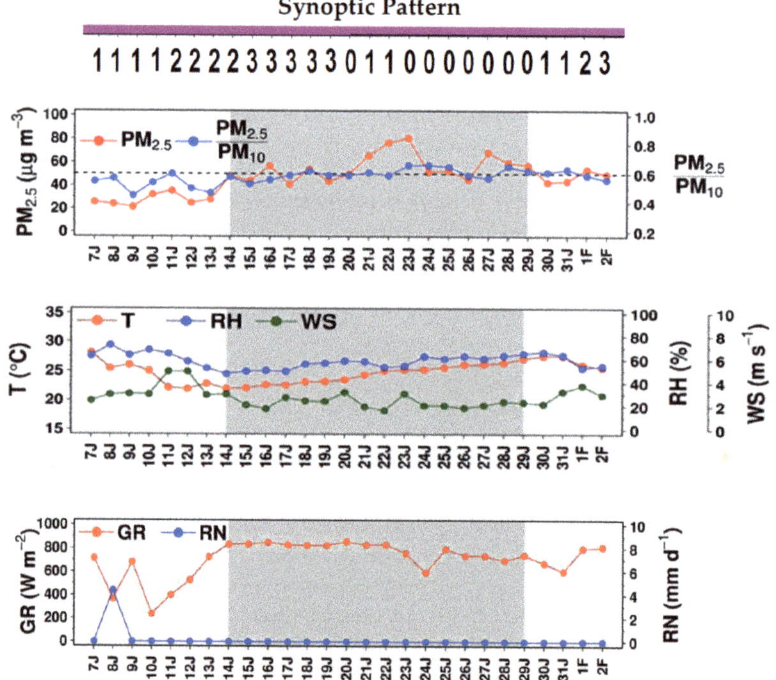

Figure 3. Daily $PM_{2.5}$, $PM_{2.5}/PM_{10}$ (at station P61) and other meteorological variables during EP1.

Global radiation was relatively low in pre-HZ1 (with the minimum of 237.1 W m^{-2}) but turns relatively large in HZ1 (with the maximum of 849.4 W m^{-2}), which could be attributed partly to fewer clouds induced by the high-pressure cold surge (i.e., shifted synoptic patterns). No rain was observed over the entire episode, except on one single day in pre-HZ1 with a light amount (4.4 mm).

In view of diurnal variation (Figure S4 in Supplementary Materials), during pre-HZ1, PM$_{2.5}$ was generally low (<30 μg m^{-3} for most of the hours). Once the haze sets in, PM$_{2.5}$ showed diurnality more clearly, i.e., relatively low in the afternoon but high in the nighttime and morning. Furthermore, we noticed that diurnal variation in the first (14–23 January) and second (24–29 January) parts of HZ1 showed a sharp contrast and PM$_{2.5}$ became relatively high in the afternoon in second part as compared to the afternoon PM$_{2.5}$ in the first part. We looked to daytime Obukhov length but did not find any dramatic change in instability over the days at all (Figure S4 in Supplementary Materials). Hence, it was not possible to explain it directly, and we then suspected that secondary aerosols were enriched in the afternoon for the later part of HZ1, given that this haze event extended as long as 16 days with increasing temperature and humidity and ample global radiation.

Figure 4 shows daily vertical temperature profiles with inversion layers identified. Low-level inversion occurred for 4 days (out of 7) during pre-HZ1 but persistently appeared almost every day (14 out of 15 days with non-missing data) in HZ1. The latter highlights the limited diluting volume for air pollutants and facilitate PM$_{2.5}$ elevation. Inversion intensity varied day-to-day during HZ1, with a minimum of 0.13 °C/100 m on 22 January and a maximum of 2.39 °C/100 m on 25 January. As mentioned previously, the temperature was relatively low after the cold surge arrives, and before it receded or dissipated, the low-level inversion was more easily induced. Even though global radiation and near-surface dynamic instability were present, it still took more heat or longer time to warm the surface to break up the inversion, based on the concept of the bulk model of daytime mixing height [47]. With a sequence of inversion-breakup failures, multiple low-level inversion layers could form, which was seen twice in HZ1. Atmospheric aerosols were known for radiative effects, e.g., black carbon or soot to absorb heat and sulfate to scatter radiation, and they could play a role in modifying temperature profiles. However, this subject is beyond the scope of the current study. Upper-level inversion also existed before the cold surge arrival and maintained over most of EP1, suggesting the presence of large-scale subsidence inversion aloft.

Lastly, the transport of air pollutants was investigated using back-trajectories (Figure 5 and Figure S5 in Supplementary Materials). Over the course of EP1, most of the back-trajectories moved from the eastern and northeastern directions, allowing traveling air masses to absorb and carry air pollutants or emissions to the study area. Coincidentally, biomass burning was present in Laos, Cambodia and the central and northeastern regions of Thailand and became relatively intense during HZ1. Only two air masses (out of 7) pass through fire areas in pre-HZ1, while all air masses move over such areas during HZ1 (Figure S6 in Supplementary Materials). Slow-moving and low-level back-trajectories found in the second part of HZ1 (24–29 January) possibly worsened the PM$_{2.5}$ situation in the study area (Figure S5 in Supplementary Materials). During post-HZ1, half of the back-trajectories (2 out of 4) were maritime (i.e., originating in or passing over the Gulf of Thailand) and thus relatively clean. It is, thus, fair to say that, besides local and synoptic meteorology, the long-range transport potentially impacted the urban haze in GBK to a certain extent in this episode.

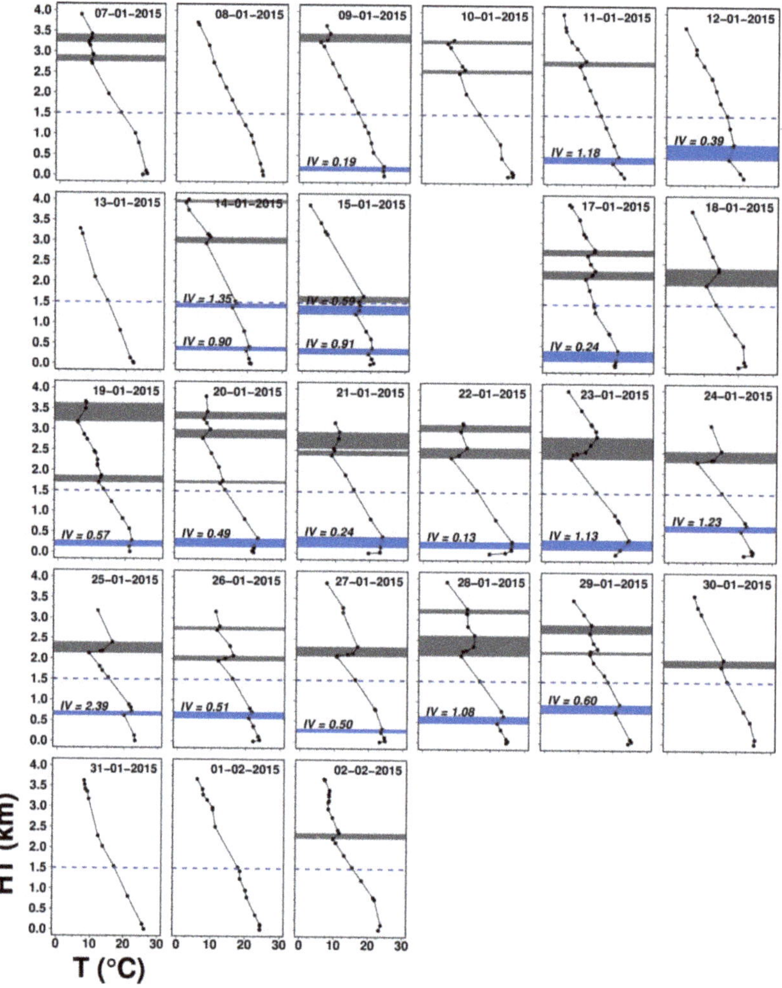

Figure 4. Daily vertical temperature profiles at 7 local time (LT) during EP1. The blue rectangles are the low-level inversion layers (found at 0.1–1.5 km), while the gray rectangles are the upper-level ones (found at 1.5–4 km). In the figure, *IV* represents the inversion intensity in °C/100 m.

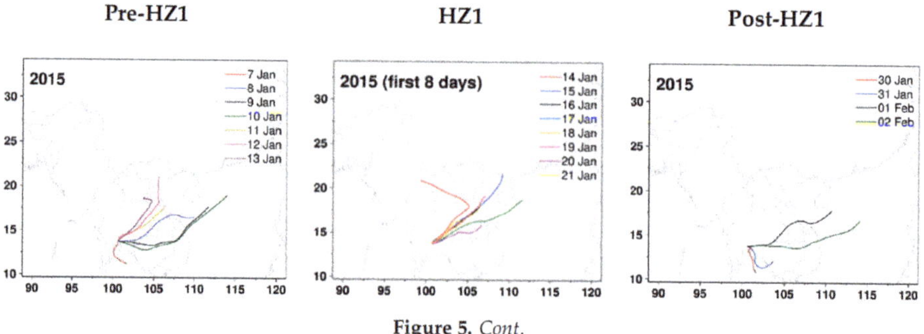

Figure 5. *Cont.*

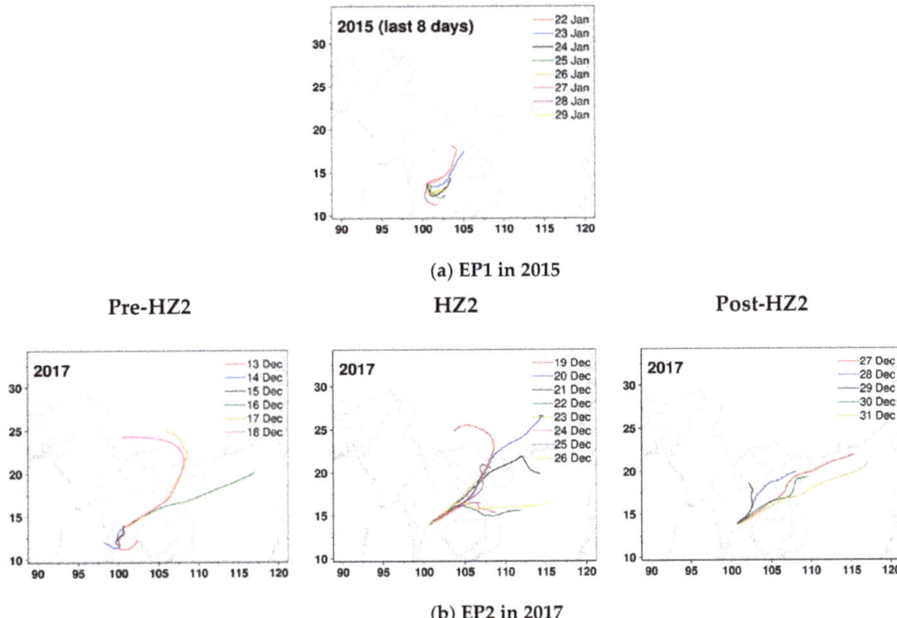

Figure 5. Daily 48-h back-trajectories: (**a**) EP1 in 2015; (**b**) EP2 in 2017.

3.2. Haze Episode EP2

This episode (EP2) occurred in December 2017, with pre-HZ2 on 13–18 December, the haze event (HZ2) on 19–26 December 2017 and post-HZ2 on 27–31 December. Daily $PM_{2.5}$, $PM_{2.5}/PM_{10}$ ratio, temperature, relative humidity, wind speed, global radiation, rain, and synoptic pattern for EP2 are given in Figure 6. As seen, the average $PM_{2.5}$ was less than 50 µg m^{-3} for all days during pre-HZ2. At the start of HZ2 on 19 December, $PM_{2.5}$ increased to 54.4 µg m^{-3} and remains consistently higher than 50 µg m^{-3} until 26 December (except on 25 December). $PM_{2.5}$ dropped sharply after 26 December due likely to wet scavenging by rain reported for three consecutive days (26–28 December) caused by a tropical depression in the Gulf of Thailand [57]. The tropical depression was the final phase of Typhoon Tembin maturely developed in the South China Sea. $PM_{2.5}/PM_{10}$ appeared to be higher in EP2 (0.69–0.94) than that in EP1 (0.45–0.66). The lower ratio in EP1 was due partly to significant contributions from agricultural burning and forest fires. Based on a global emission database by Klimont et al. (2017) [58], it was found that the $PM_{2.5}/PM_{10}$ ratio for emissions from agricultural burning and forest fires was lower as compared to those for other anthropogenic emissions. Thus, biomass burning observed in EP1 may have decreased the $PM_{2.5}/PM_{10}$ ratio. Haze episode EP2 did not show effects of biomass burning, which is explained in the latter part of this section. Temperature and relative humidity on 13 December was 30 °C and 74.1%, respectively and did not change much until 16 December. The synoptic change from Pattern 1 to Pattern 2 on 16–17 December, which caused an abrupt decrease in both variables until Pattern 2 prevailed, i.e., 20 December. Temperature and relative humidity dropped to 20.5 °C and 45.1%, respectively, on 20 December and gradually climbed up to 28.8 °C and 56.4%, respectively, on 24 December due to the weakening of cold surge leading to the return of the warm moist air. Once the cold surge recedes, Pattern 0 and Pattern 1 were dominant for most of the days. Global radiation increased from 393.8 W m^{-2} on 13 December to 594.7 W m^{-2} on 19 December (i.e., the start of HZ2) due to less cloud cover caused by the cold surge. Over 24–27 December, both temperature and global radiation declined and relative humidity increased, as influenced by the tropical depression. The tropical depression and rainfall observed during 26–28 December likely increased relative humidity during 24–27 December. The temperature rose after 27 December until the end of EP2 due to an increase in global radiation, which in turn decreased relative humidity. Global radiation was consistently higher during the haze event until 24 December with less variation and decreased later between 24 and 27 December and then again increased until the end of EP2 during 27–31 December. The wind followed the synoptic patterns quite closely. It was relatively strong during the active phase of cold surge during pre-HZ2 (particularly, 19–22 December) and starting four days of HZ2 while for the other days it was relatively weak.

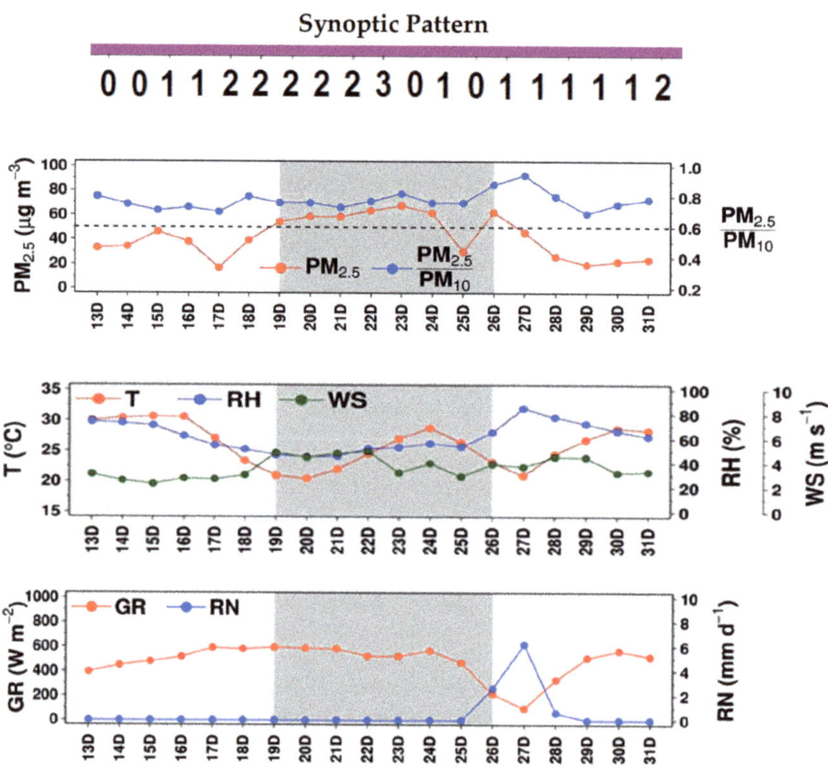

Figure 6. Daily $PM_{2.5}$, $PM_{2.5}/PM_{10}$, and other meteorological variables during EP2.

The diurnality during HZ2 was clearer as compared to that during the pre-HZ2 and post-HZ2 (Figure S4 in Supplementary Materials). An obvious diurnal pattern with higher $PM_{2.5}$ during the night and early morning and drop during the afternoon were observed during HZ2. Obukhov length indicated stability conditions for some hours during a few days. However, no clear change in stability was observed throughout the haze event before the tropical depression.

The vertical temperature profiles with inversion layers for all days in this episode are shown in Figure 7. Three out of six days during pre-HZ2 do not show any temperature inversion while all non-missing seven days during HZ2 between 19 and 26 December show temperature inversions of varying intensity strength with a minimum of 0.25 °C/100 m on 26 December and a maximum of 2.0 °C/100 m on 24 December (Figure 7). Multiple low-level inversion layers were found during 21–22 December. 2 out of 5 days during post-HZ2 did not show any inversion. High-level inversions existed for most of the days during EP2 due to the presence of large-scale subsidence inversion aloft in the winter.

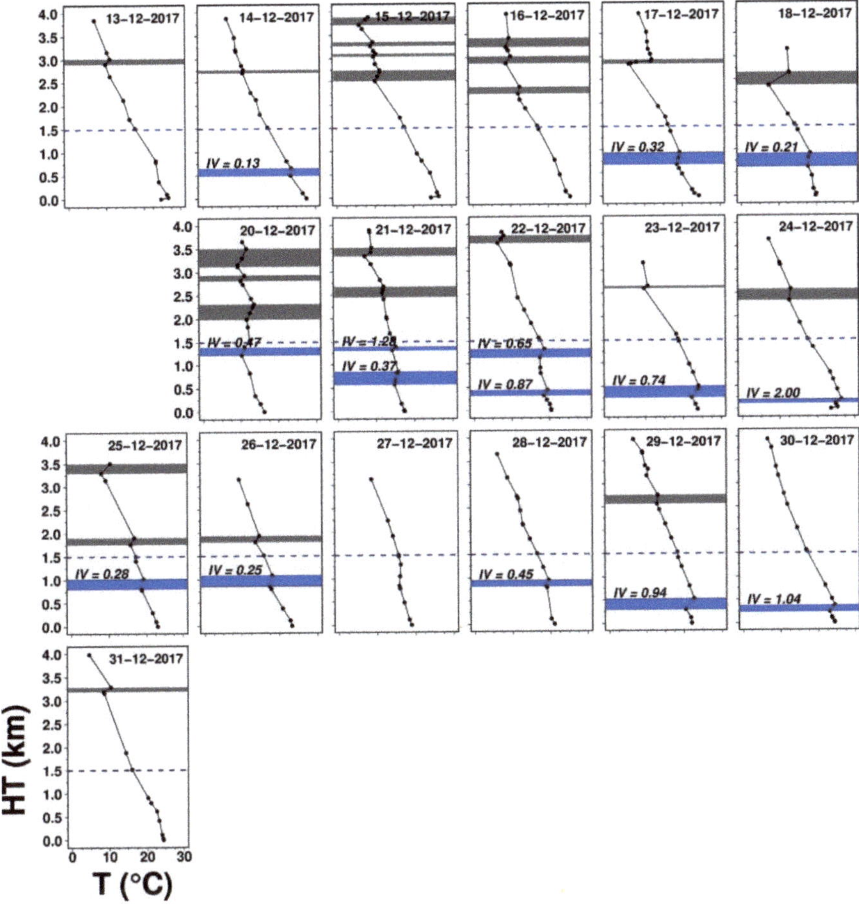

Figure 7. Same as Figure 4, but for haze episode EP2. *IV* represents the inversion intensity in °C/100 m.

Lastly, we used back-trajectories from HYSPLIT to investigate the effect of long-range transport of air pollutants (Figure 5 and Figure S5 in Supplementary Materials). Except for 13–14 December, all the days had longer trajectories and coming from the eastern and northeastern directions. Trajectories during pre-HZ2 and post-HZ2 were mostly low level, while for HZ2, it was at a high level for many days (Figure S5 in Supplementary Materials). Fewer fire hot spots for all three periods: pre-HZ2, HZ2, and post-HZ2 were found in nearby regions suggesting that this haze episode HZ2 was mainly caused by the synergetic effect of local emission and local/synoptic meteorology (Figure S6 in Supplementary Materials). Comparing the two episodes, the EP2 back-trajectories tended to be longer (i.e., move faster). Fast-moving back-trajectories generally had short residential times to absorb atmospheric constituents. These imply limited contribution from biomass burning to haze, and elevated haze in this episode may have been more driven by local emissions.

It is noted that the contrasting results of the potential impact of biomass burning on the two episodes suggest that the influence of biomass burning may have varied by episode. Following our literature survey for Bangkok or Greater Bangkok (see Table S1 in Supplementary Materials), Phairuang et al. (2019) [33] found agricultural burning in the central region of Thailand and forest fires in remote areas (e.g., the northern region) to contribute to particulate matter in Bangkok, but the impact is more evident during the winter (November–January) when air masses are generally continental and favorably transporting haze to the study area, as compared to the summer (March) when air masses are relatively clean due to originating in or passing over the Gulf of Thailand. Similarly, Kayee et al. (2020) [35] looked into a summertime haze episode and found no significant contribution from biomass burning in the northern region due likely to lack of continental air masses. Dejchanchaiwong et al. (2020) [34] reported biomass burning in Thailand and Cambodia with favorable air masses as a major contributor to a wintertime haze episode.

4. Conclusions

Several local and synoptic meteorological aspects associated with the selected two wintertime haze episodes in Greater Bangkok were examined using observational data, including classified four synoptic patterns, day-to-day and diurnal variation, dynamic stability, temperature inversion, and back-trajectories. The first episode included an elevated haze event of 16 days (14–29 January 2015), together with some days before and after the haze event. However, the second episode had an elevated haze event of only 8 days (19–26 December 2017). Daily $PM_{2.5}$ was found to be 50 µg m^{-3} or higher over most of each haze event. The two haze events commonly had cold surges as the background synoptic feature to initiate or trigger haze evolution. The cold surge reached the study area before the start of each haze event, causing temperature and relative humidity to drop abruptly initially but then to gradually increase as the cold surge weakened or dissipated. Wind speed was relatively large when the cold surge was active. Global radiation was generally modulated by cloud cover, which turns relatively high during each haze event because the cold surge induced fewer clouds. Daytime dynamic stability was generally unstable along the course of each haze event, except being stable at the ending of the second haze event due to a tropical depression. In each haze event, low-level temperature inversion existed well, with double layers seen in the beginning, effectively suppressing atmospheric dilution. Large-scale subsidence inversion aloft was also persistently present. In the first haze event, relatively low $PM_{2.5}$ was observed in the afternoon over its first part, but a higher afternoon $PM_{2.5}$ occurred over its second part, due possibly to the role of secondary aerosols. Comparing the two episodes, the $PM_{2.5}/PM_{10}$ ratio was relatively low in the first episode because of more impact of biomass burning, which was in general agreement with back-trajectories and active fire hotspots. The second haze event, with little biomass burning in the region, was likely to be caused mainly by local anthropogenic emissions.

According to the above findings, certain policy-related implications are given as follows:

(1) Local and synoptic meteorological factors play an important role in the elevated haze over Greater Bangkok. This poses a challenge to haze mitigation by emission reduction. Cold surge leads to an abrupt change in local meteorological conditions with less atmospheric ventilation as a result. A time window to apply control measures for haze is thus limited, emphasizing readiness in haze mitigation responses and early public warning with reliable weather/air quality forecasting.

(2) Apart from conventional anthropogenic emissions, biomass burning is another emission sector that potentially contributes to the urban haze in the study area. Thus, agricultural burning and forest fires need to be considered, and better controlled/managed in support of haze mitigation.

(3) It is fair to say that the contributing degrees of different emission sectors (i.e., anthropogenic and biomass burning) may vary by episode or over days within an episode and that secondary aerosols can as well form and partly contribute. Haze-related policymaking needs an integrated approach dealing with all important emission sectors for both particulate and gaseous precursors.

Future works on the urban haze problem in the study area may be extended to the following aspects: urban heat island, local recirculation, and systematic synoptic pattern classification [59], hourly mixing height variation and application of high-resolution light detection and ranging (LiDAR) if available [60], low-visibility events, and episode-specific source–receptor studies using source apportionment, chemical characterization (e.g., carbonaceous components in $PM_{2.5}$), and air quality modeling.

Supplementary Materials: The following are available online at http://www.mdpi.com/1660-4601/17/24/9499/s1. Table S1. Selected air pollution studies in Greater Bangkok. Figure S1. Daily $PM_{2.5}$ variation at the three air quality stations by year: (a) 2015; (b) 2016; (c) 2017. Figure S2. Example surface charts in EP1 corresponding to the four synoptic patterns classified: (a) Pattern 0; (b) Pattern 1; (c) Pattern 2; (d) Pattern 3. Figure S3. Example surface charts in EP2 corresponding to the four synoptic patterns classified: (a) Pattern 0; (b) Pattern 1; (c) Pattern 2; (d) Pattern 3. Figure S4. Diurnal variation of $PM_{2.5}$ and Obukhov length (L) during: (a) EP1 in 2015; (b) EP2 in 2017. Figure S5. Vertical migration of each individual daily 48-h back-trajectory: (a) EP1 in 2015; (b) EP2 in 2017. Figure S6. Average daily active fire hotspots per pixel (0.5° by 0.5°): (a) EP1 in 2015; (b) EP2 in 2017.

Author Contributions: K.M. initiated, conceptualized, and administrated the research. N.A. performed the entire analysis and drafted the manuscript. Both K.M. and N.A. jointly discussed the results. N.P.-E., E.K., T.B., S.P., and C.C. technically supported the research. B.D. contributed to the data processing. All authors have read and agreed to the published version of the manuscript.

Funding: This research was funded mainly by the Energy Conservation and Promotion Fund (ENCONFUND) of the Ministry of Energy, Thailand, and partly by the Joint Graduate School of Energy and Environment (JGSEE) and the Postgraduate Education and Research Development Office (PERDO).

Acknowledgments: The authors sincerely thank the Pollution Control Department (PCD) and the Thai Meteorological Department (TMD) for the observational data used in the study. Seksan Sangkow (PCD) and Thiantawan Chulathipyachat (PCD) shared useful perspectives on air quality issues in Greater Bangkok and the use of air quality data. We also thank the two anonymous reviewers for the comments and suggestions that helped improve the manuscript.

Conflicts of Interest: The authors declare no conflict of interest.

References

1. Zhang, R.; Wang, G.; Guo, S.; Zamora, M.L.; Ying, Q.; Lin, Y.; Wang, W.; Hu, M.; Wang, Y. Formation of urban fine particulate matter. *Chem. Rev.* **2015**, *115*, 3803–3855. [CrossRef] [PubMed]

2. Seinfeld, J.H.; Pandis, S.N. *Atmospheric Chemistry and Physics from Air Pollution to Climate Change*, 2nd ed.; John Wiley & Sons, Inc.: New York, NY, USA, 2006; ISBN 9780471720171.

3. Health Effects Institute (HEI). *State of Global Air 2017*; A Special Report on Global Exposure to Air Pollution and Its Disease Burden; Health Effects Institute, The Institute for Health Metrics and Evaluation, University of British Columbia: Boston, MA, USA, 2018; Available online: https://www.stateofglobalair.org/sites/default/files/soga_2017_report.pdf (accessed on 15 September 2020).

4. Sun, Y.; Chen, C.; Zhang, Y.; Xu, W.; Zhou, L.; Cheng, X.; Zheng, H.; Ji, D.; Li, J.; Tang, X.; et al. Rapid formation and evolution of an extreme haze episode in Northern China during winter 2015. *Sci. Rep.* **2016**, *6*, 27151. [CrossRef] [PubMed]

5. Song, M.; Liu, X.; Tan, Q.; Feng, M.; Qu, Y.; An, J. Characteristics and formation mechanism of persistent extreme haze pollution events in Chengdu, southwestern China. *Environ. Pollut.* **2019**, *251*, 1–12. [CrossRef] [PubMed]

6. Chen, L.; Zhu, J.; Liao, H.; Gao, Y.; Qiu, Y.; Zhang, M.; Liu, Z.; Li, N.; Wang, Y. Assessing the formation and evolution mechanisms of severe haze pollution in the Beijing–Tianjin–Hebei region using process analysis. *Atmos. Chem. Phys.* **2019**, *19*, 10845–10864. [CrossRef]

7. Dai, Z.; Liu, D.; Yu, K.; Cao, L.; Jiang, Y. Meteorological variables and synoptic patterns associated with air pollutions in eastern China during 2013–2018. *Int. J. Environ. Res. Public Health* **2020**, *17*, 2528. [CrossRef] [PubMed]

8. Kanawade, V.P.; Srivastava, A.K.; Ram, K.; Asmi, E.; Vakkari, V.; Soni, V.K.; Varaprasad, V.; Sarangi, C. What caused severe air pollution episode of November 2016 in New Delhi? *Atmos. Environ.* **2020**, *222*, 117125. [CrossRef]

9. Wang, S.; Liao, T.; Wang, L.; Sun, Y. Process analysis of characteristics of the boundary layer during a heavy haze pollution episode in an inland megacity, China. *J. Environ. Sci.* **2016**, *40*, 138–144. [CrossRef]

10. Guo, S.; Hu, M.; Zamora, M.L.; Peng, J.; Shang, D.; Zheng, J.; Du, Z.; Wu, Z.; Shao, M.; Zeng, L.; et al. Elucidating severe urban haze formation in China. *Proc. Natl. Acad. Sci. USA* **2014**, *111*, 17373–17378. [CrossRef]

11. Yu, S.; Li, P.; Wang, L.; Wu, Y.; Wang, S.; Liu, K.; Zhu, T.; Zhang, Y.; Hu, M.; Zeng, L.; et al. Mitigation of severe urban haze pollution by a precision air pollution control approach. *Sci. Rep.* **2018**, *8*, 41598. [CrossRef]

12. Mishra, D.; Goyal, P.; Upadhyay, A. Artificial intelligence based approach to forecast $PM_{2.5}$ during haze episodes: A case study of Delhi, India. *Atmos. Environ.* **2015**, *102*, 239–248. [CrossRef]
13. Huang, R.J.; Zhang, Y.; Bozzetti, C.; Ho, K.F.; Cao, J.J.; Han, Y.; Daellenbach, K.R.; Slowik, J.G.; Platt, S.M.; Canonaco, F.; et al. High secondary aerosol contribution to particulate pollution during haze events in China. *Nature* **2015**, *514*, 218–222. [CrossRef] [PubMed]
14. Quan, J.; Tie, X.; Zhang, Q.; Liu, Q.; Li, X.; Gao, Y.; Zhao, D. Characteristics of heavy aerosol pollution during the 2012–2013 winter in Beijing, China. *Atmos. Environ.* **2014**, *88*, 83–89. [CrossRef]
15. Zhang, Y.L.; Cao, F. Fine particulate matter ($PM_{2.5}$) in China at a city level. *Sci. Rep.* **2015**, *5*, 14884. [CrossRef]
16. Zhang, Q.; Quan, J.; Tie, X.; Li, X.; Liu, Q.; Gao, Y.; Zhao, D. Effects of meteorology and secondary particle formation on visibility during heavy haze events in Beijing, China. *Sci. Total Environ.* **2015**, *502*, 578–584. [CrossRef]
17. Mao, M.; Zhang, X.; Shao, Y.; Yin, Y. Spatiotemporal variations and factors of air quality in urban central China during 2013–2015. *Int. J. Environ. Res. Public Health* **2019**, *17*, 229. [CrossRef] [PubMed]
18. Van Ulden, A.P.; Holtslag, A.A.M. Estimation of atmospheric boundary layer parameters for diffusion applications. *J. Clim. Appl. Meteor.* **1985**, *24*, 1196–1207. [CrossRef]
19. Quan, J.; Gao, Y.; Zhang, Q.; Tie, X.; Cao, J.; Han, S.; Meng, J.; Chen, P.; Zhao, D. Evolution of planetary boundary layer under different weather conditions, and its impact on aerosol concentrations. *Particuology* **2013**, *11*, 34–40. [CrossRef]
20. Petaja, T.; Jarvi, L.; Kerminen, V.M.; Ding, A.J.; Sun, J.N.; Nie, W.; Kujansuu, J.; Virkkula, A.; Yang, X.Q.; Fu, C.B.; et al. Enhanced air pollution via aerosol-boundary layer feedback in China. *Sci. Rep.* **2015**, *6*, 18998. [CrossRef]
21. Tie, X.; Huang, R.J.; Cao, J.; Zhang, Q.; Cheng, Y.; Su, H.; Chang, D.; Pöschl, U.; Hoffmann, T.; Dusek, U.; et al. Severe pollution in China amplified by atmospheric moisture. *Sci. Rep.* **2017**, *7*, 15760. [CrossRef]
22. Liu, Q.; Jia, X.; Quan, J.; Li, J.; Li, X.; Wu, Y.; Chen, D.; Wang, Z.; Liu, Y. New positive feedback mechanism between boundary layer meteorology and secondary aerosol formation during severe haze events. *Sci. Rep.* **2018**, *8*, 41598. [CrossRef]
23. Li, X.; Wang, Y.; Zhao, H.; Hong, Y.; Liu, N.; Ma, Y. Characteristics of pollutants and boundary layer structure during two haze events in summer and autumn 2014 in Shenyang, northeast China. *Aerosol. Air Qual. Res.* **2018**, *18*, 386–396. [CrossRef]
24. Li, X.; Wang, Y.; Shen, L.; Zhang, H.; Zhao, H.; Zhang, Y.; Ma, Y. Characteristics of boundary layer structure during a persistent haze event in the central Liaoning city cluster, northeast China. *J. Meteorol. Res.* **2018**, *32*, 302–312. [CrossRef]
25. Huaqing, P.; Duanyang, L.; Bin, Z.; Yan, S.; Jiamei, W.; Hao, S.; Jiansu, W.; Lu, C. Boundary-layer characteristics of persistent regional haze events and heavy haze days in eastern China. *Adv. Meteorol.* **2015**, *2016*. [CrossRef]
26. Li, Q.; Wu, B.; Liu, J.; Zhang, H.; Cai, X.; Song, Y. Characteristics of the atmospheric boundary layer and its relation with $PM_{2.5}$ during haze episodes in winter in the north China plain. *Atmos. Environ.* **2020**, *223*, 117265. [CrossRef]
27. Ye, X.; Song, Y.; Cai, X.; Zhang, H. Study on the synoptic flow patterns and boundary layer process of the severe haze events over the north China plain in January 2013. *Atmos. Environ.* **2016**, *124*, 129–145. [CrossRef]
28. Pollution Control Department (PCD). *Thailand State of Pollution 2017*; Pollution Control Department: Bangkok, Thailand, 2018.
29. Chuersuwan, N.; Nimrat, S.; Lekphet, S.; Kerdkumrai, T. Levels and major sources of $PM_{2.5}$ and PM_{10} in Bangkok Metropolitan Region. *Environ. Int.* **2008**, *34*, 671–677. [CrossRef]
30. Wimolwattanapun, W.; Hopke, P.K.; Pongkiatkul, P. Source apportionment and potential source locations of $PM_{2.5}$ and $PM_{2.5-10}$ at residential sites in metropolitan Bangkok. *Atmos. Pollut. Res.* **2011**, *2*, 172–181. [CrossRef]
31. ChooChuay, C.; Pongpiachan, S.; Tipmanee, D.; Suttinun, O.; Deelaman, W.; Wang, Q.; Xing, L.; Li, G.; Han, Y.; Palakun, J.; et al. Impacts of $PM_{2.5}$ sources on variations in particulate chemical compounds in ambient air of Bangkok, Thailand. *Atmos. Pollut. Res.* **2020**, *11*, 1657–1667. [CrossRef]
32. Narita, D.; Oanh, N.T.K.; Sato, K.; Huo, M.; Permadi, D.A.; Chi, N.N.H.; Ratanajaratroj, T.; Pawarmart, I. Pollution characteristics and policy actions on fine particulate matter in a growing Asian economy: The case of Bangkok Metropolitan Region. *Atmosphere* **2019**, *10*, 227. [CrossRef]

33. Phairuang, W.; Suwattiga, P.; Chetiyanukornkul, T.; Hongtieab, S.; Limpaseni, W.; Ikemori, F.; Hata, M.; Furuuchi, M. The influence of the open burning of agricultural biomass and forest fires in Thailand on the carbonaceous components in size-fractionated particles. *Environ. Pollut.* **2019**, *247*, 238–247. [CrossRef]
34. Dejchanchaiwong, R.; Tekasakul, P.; Tekasakul, S.; Phairuang, W.; Nim, N.; Sresawasd, C.; Thongboon, K.; Thongyen, T.; Suwattiga, P. Impact of transport of fine and ultrafine particles from open biomass burning on air quality during 2019 Bangkok haze episode. *J. Environ. Sci.* **2020**, *97*, 238–247. [CrossRef] [PubMed]
35. Kayee, J.; Sompongchaiyakul, P.; Sanwlani, N.; Bureekul, S.; Wang, X.; Das, R. Metal concentrations and source apportionment of $PM_{2.5}$ in Chiang Rai and Bangkok, Thailand during a biomass burning season. *ACS Earth Space Chem.* **2020**, *4*, 1213–1226. [CrossRef]
36. Pham, T.B.T.; Manomaiphiboon, K.; Vongmahadlek, C. Development of an inventory and temporal allocation profiles of emissions from power plants and industrial facilities in Thailand. *Sci. Total Environ.* **2008**, *397*, 103–118. [CrossRef] [PubMed]
37. Thongthammachart, T.; Jinsart, W. Estimating $PM_{2.5}$ concentrations with statistical distribution techniques for health risk assessment in Bangkok. *Hum. Ecol. Risk Assess.* **2020**, *26*, 1848–1863. [CrossRef]
38. Fold, N.R.; Allison, M.R.; Wood, B.C.; Pham, T.B.T.; Bonnet, S.; Garivait, S.; Kamens, R.; Pengjan, S. An assessment of annual mortality attributable to ambient $PM_{2.5}$ in Bangkok, Thailand. *Int. J. Environ. Res. Public Health* **2020**, *17*, 7298. [CrossRef]
39. DOPA–Department of Provincial Administration. Statistic of Population by Province in 2017. Available online: https://stat.bora.dopa.go.th/stat/pk/pk_60.pdf (accessed on 28 September 2020).
40. National Economic and Social Development Board (NESDB). *Gross Regional and Provincial Product, Chain Volume Measures*, 2018th ed.; Office of the National Economic and Social Development Board: Bangkok, Thailand, 2018. Available online: https://www.nesdc.go.th/nesdb_en/ewt_w3c/ewt_dl_link.php?filename=national_account&nid=4317 (accessed on 28 September 2020).
41. Land Development Department (LDD). *Land Use and Land Cover Data for Thailand for the Years 2012–2016*; Land Development Department: Bangkok, Thailand, 2016.
42. Thai Meteorological Department (TMD). The Climate of Thailand. Thai Meteorological Department. Available online: https://www.tmd.go.th/en/archive/thailand_climate.pdf (accessed on 28 September 2020).
43. Chang, C.P.; Harr, P.A.; Chen, H.J. Synoptic disturbances over the equatorial south China sea and western maritime continent during boreal winter. *Mon. Weather Rev.* **2005**, *133*, 489–503. [CrossRef]
44. Ashfold, M.J.; Latif, M.T.; Samah, A.A.; Mead, M.I.; Harris, N.R.P. Influence of northeast monsoon cold surges on air quality in Southeast Asia. *Atmos. Environ.* **2017**, *166*, 498–509. [CrossRef]
45. Yavinchan, S.; Exell, R.H.B.; Sukawat, D. Convective parameterization in a model for prediction of heavy rain in southern Thailand. *J. Meteor. Soc. Jpn.* **2011**, *89A*, 201–224. [CrossRef]
46. Zahumenský, I. Guidelines on Quality Control Procedures for Data from Automatic Weather Stations. World Meteorolgical Organziation. 2004. Available online: https://www.wmo.int/pages/prog/www/IMOP/meetings/Surface/ET-STMT1_Geneva2004/Doc6.1(2).pdf (accessed on 13 February 2019).
47. Aman, N.; Manomaiphiboon, K.; Pengchai, P.; Suwanathada, P.; Srichawana, J.; Assareh, N. Long-term observed visibility in eastern Thailand: Temporal variation, association with air pollutants and meteorological factors, and trends. *Atmosphere* **2019**, *10*, 122. [CrossRef]
48. R Development Core Team. *R: A Language and Environment for Statistical Computing (Version 3.6.1)*; R Foundation for Statistical Computing: Vienna, Austria, 2019; Available online: https://www.R-project.org/ (accessed on 20 June 2020).
49. Saha, S.; Moorthi, S.; Pan, H.L.; Wu, X.; Wang, J.; Nadiga, S.; Tripp, P.; Kistler, R.; Woollen, J.; Behringer, D.; et al. The NCEP climate forecast system reanalysis. *Bull. Am. Meteorol. Soc.* **2010**, *91*, 1015–1057. [CrossRef]
50. Stull, R.B. *An Introduction to Boundary Layer Meteorology*; Kluwer Academic Publishers: Dordrecht, The Netherlands, 1998.
51. Kamma, J.; Manomaiphiboon, K.; Aman, N.; Thongkamdee, T.; Chuangchote, S.; Bonnet, S. Urban heat island analysis for Bangkok: Multi-scale temporal variation, associated factors, directional dependence, and cool island condition. *ScienceAsia* **2020**, *46*, 213–223. [CrossRef]
52. Stewart, I.D.; Oke, T.R. Local climate zones for urban temperature studies. *Bull. Am. Meteorol. Soc.* **2012**, *12*, 1879–1900. [CrossRef]

53. Sáenz, J.; González-Rojí, S.J.; Madinabeitia, S.C.; Berastegi, G.I. Atmospheric Thermodynamics and Visualization. R Package Version 1.2.1. 2018. Available online: https://cran.r-project.org/web/packages/aiRthermo/index.html (accessed on 20 June 2020).

54. Rodell, M.; Houser, P.R.; Jambor, U.; Gottschalck, J.; Mitchell, K.; Meng, C.J.; Arsenault, K.; Cosgrove, B.; Radakovich, J.; Bosilovich, M.; et al. The global land data assimilation system. *Bull. Am. Meteor. Soc.* **2004**, *85*, 381–394. [CrossRef]

55. Stein, A.F.; Draxler, R.R.; Rolph, G.D.; Stunder, B.J.B.; Cohen, M.D.; Ngan, F. NOAA's Hysplit atmospheric transport and dispersion modeling system. *Bull. Am. Meteorol. Soc.* **2015**, *96*, 2059–2077. [CrossRef]

56. Giglio, L.; Schroeder, W.; Justice, C.O. The collection 6 MODIS active fire detection algorithm and fire products. *Remote Sens. Environ.* **2016**, *178*, 31–41. [CrossRef]

57. Japan Meteorological Agency (JMA). *Annual Report on the Activities of RSMC Tokyo*; Typhoon Center: Tokyo, Japan, 2018. Available online: https://www.jma.go.jp/jma/jma-eng/jma-center/rsmc-hp-pub-eg/AnnualReport/2017/Text/Text2017.pdf (accessed on 28 September 2020).

58. Klimont, Z.; Kupiainen, K.; Heyes, C.; Purohit, P.; Cofala, J.; Rafaj, P.; Borken-Kleefeld, J.; Schöpp, W. Global anthropogenic emissions of particulate matter including black carbon. *Atmos. Chem. Phys.* **2017**, *17*, 8681–8723. [CrossRef]

59. Pérez, I.A.; García, M.Á.; Sánchez, M.L.; Pardo, N.; Fernández-Duque, B. Key points in air pollution meteorology. *Int. J. Environ. Res. Public Health* **2020**, *17*, 8349. [CrossRef]

60. Solanki, R.; Macatangay, R.; Sakulsupich, V.; Sonkaew, T.; Sarathi, P. Mixing layer height retrievals from MiniMPL measurements in the Chiang Mai valley: Implications for particulate matter pollution. *Front. Earth Sci.* **2019**, *7*. [CrossRef]

Publisher's Note: MDPI stays neutral with regard to jurisdictional claims in published maps and institutional affiliations.

International Journal of
**Environmental Research
and Public Health**

Article

Meteorological Variables and Synoptic Patterns Associated with Air Pollutions in Eastern China during 2013–2018

Zhujun Dai [1,2], Duanyang Liu [1,3,4,](#) , Kun Yu [2], Lu Cao [5] and Youshan Jiang [2]

1   Key Laboratory of Transportation Meteorology, China Meteorological Administration, Nanjing 210008, China; daizhujun99@163.com
2   Nanjing Meteorological Bureau of Jiangsu Province, Nanjing 210019, China; mitsuiyk@163.com (K.Y.); jysnjsqxt@163.com (Y.J.)
3   Jiangsu Institute of Meteorological Sciences, Nanjing 210008, China
4   Nanjing Joint Institute for Atmospheric Sciences, Nanjing 210008, China
5   Meteorological Observatory of Jiangsu Province, Nanjing 210008, China; caolu36@163.com
*   Correspondence: liuduanyang2001@126.com; Tel.: +0086-18168100412

Received: 24 February 2020; Accepted: 4 April 2020; Published: 7 April 2020

Abstract: Steady meteorological conditions are important external factors affecting air pollution. In order to analyze how adverse meteorological variables affect air pollution, surface synoptic situation patterns and meteorological conditions during heavy pollution episodes are discussed. The results showed that there were 78 RPHPDs (regional $PM_{2.5}$ pollution days) in Jiangsu, with a decreasing trend year by year. Winter had the most stable meteorological conditions, thus most RPHPDs appeared in winter, followed by autumn and summer, with the least days in spring. RPHPDs were classified into three patterns, respectively, as equalized pressure (EQP), advancing edge of a cold front (ACF) and inverted trough of low pressure (INT) according to the SLP (sea level pressure). RPHPDs under EQP were the most (51%), followed by ACF (37%); INT was the minimum (12%). Using statistical methods and meteorological condition data on RPHPDs from 2013 to 2017 to deduce the thresholds and 2018 as an independent dataset to validate the proposed thresholds, the threshold values of meteorological elements are summarized as follows. The probability of RPHPDs without rain was above 92% with the daily and hourly precipitation of all RPHPDs below 2.1 mm and 0.8 mm. Wind speed, RHs, inversion intensity(ITI), height difference in the temperature inversion(ITK), the lower height of temperature inversion (LHTI) and mixed-layer height (MLH) in terms of 25%–75% high probability range were respectively within 0.5–3.6 m s^{-1}, 55%–92%, 0.7–4.0 °C 100 m $^{-1}$, 42–576 m, 3–570 m, 200–1200 m. Two conditions should be considered: whether the pattern was EQP, ACF or INT and whether the eight meteorological elements are within the thresholds. If both criteria are met, $PM_{2.5}$ particles tend to accumulate and air pollution diffusion conditions are poor. Unfavorable meteorological conditions are the necessary, but not sufficient condition for RPHPDs.

Keywords: air pollution; synoptic situation pattern; meteorological variables; threshold values

---

1. Introduction

$PM_{2.5}$ is the air particulate matter with an aerodynamically equivalent diameter of less than 2.5 μm [1]. Rising $PM_{2.5}$ levels can lead to worsening air quality, seriously affecting human health. Epidemiological studies have shown associations between ambient air pollution and changes in heart rate variability (HRV) [2–5]. $PM_{2.5}$ is characterized by the small size and lightweight. It can be transported to some faraway places from its source region, causing a wide range of heavy air pollution [6,7], leading to a wide range of human health hazards [8,9]. As a typical pollutant in heavy pollution weather, $PM_{2.5}$ has become a hot spot of research recently [10–13]. Beijing–Tianjin–Hebei region [14–17], the Yangtze River Delta region [18–22] and the Pearl River Delta region [23–25] are the three highly air-polluted regions in China. As a result, the incidence of air pollution-related diseases in major cities in these regions has been increasing year by year [8,9,26,27].

Although air pollution is usually linked with human activities, natural processes may also determine noticeable concentrations of hazardous substances in the low atmosphere. Many studies have suggested that continuous air polluting processes are jointly affected by pollutant emissions and the weather conditions that favor the accumulation of pollutants [22,23,28]. Comprehensive analyses of meteorological factors on air pollution were conducted by Zhou [29], who established the statistical model of $PM_{2.5}$ concentrations and meteorological element field. The levels of pollutants may be reduced when emissions can be controlled. However, the impact of meteorological variables on concentrations measured may be marked, and these variables cannot be controlled. If pollutant emission is the internal factor of heavy air pollution, then the meteorological condition should be the external factor [30,31]. Deducing the threshold values of meteorological elements that are conducive to the pollution accumulation is very necessary.

The Yangtze River Delta where Jiangsu Province is located is one of China's most economically developed regions—and one of the most polluted regions. In this region, many questions have been studied by scientists, such as how meteorological conditions including wind speed affect the concentration of the atmospheric pollutants, how the air pollutants transported from other heavily polluted areas by the synoptic situation patterns and meteorological variables and how changes of atmospheric mixed-layer and temperature inversions affect the air quality. Since most of the past studies are cases studies, they have not provided systematic and comprehensive answers. Moreover, the thresholds of various meteorological elements are not given. Therefore, in order to in-depth knowledge and systematic study the different synoptic situation patterns and meteorological variables on air pollution, heavy air pollution from 2013 to 2018 in Jiangsu Province, China was analyzed in this article. According to the sea level pressure (SLP), the characteristics of the boundary meteorological element distribution are discussed in different seasons, different synoptic patterns and meteorological variables.

The remainder of this paper is organized as follows: Data and methods are described in Section 2. We analyzed synoptic patterns and meteorological conditions associated with heavy $PM_{2.5}$ air pollutions in Section 3. The conclusions are given in Section 4.

2. Materials and Methods

2.1. Study Area and Data Description

The geographical location of Jiangsu Province, China (UTC/GMT+08:00), the distribution of meteorological stations and state-controlled environmental protection stations (SCEPSs) in 13 prefecture-level cities in Jiangsu were shown in Figure 1. In regards to the issue of mismatch between SCEPSs and meteorological stations, the data of SCEPSs nearest to 13 meteorological stations was adopted. According to diurnal variation characteristics and regional differences of heavy $PM_{2.5}$ pollution, 13 cities in Jiangsu Province were divided into four regions: South Jiangsu (Suzhou, Wuxi and Changzhou), Coastal Jiangsu (Lianyungang, Yancheng and Nantong), Southwest Jiangsu (Zhenjiang, Yangzhou, Taizhou and Nanjing) and North Jiangsu (Xuzhou, Huaian and Suqian). Although SCEPSs were distributed in various environments (cities, suburbs, roadsides, parks, etc.), this paper focused on the regional $PM_{2.5}$ heavy pollution, therefore the environmental differences were ignored.

The $PM_{2.5}$ data of Jiangsu Province begins from 2013, therefore the data from 2013–2018 were used in this paper. Meteorological characteristics were discussed by the data from 2013 to 2017 and verified by the data of 2018. The dataset consists of four parts: (1) routine meteorological observation data including the hourly wind speed, wind direction, surface relative humidity (RHs) and precipitation at 13 meteorological stations distributed separately in prefecture-level cities in Jiangsu from 2013 to 2018; (2) the vertical distribution of temperature was obtained from the observational data from four radiosonde stations(Xuzhou, Sheyang, Nanjing and Baoshan) at 8 o'clock. Jiangsu just had three radiosondes, which were respectively Xuzhou, Sheyang, Nanjing, located in the North, Coastal, Southwest Jiangsu. Baoshan, the nearest radiosonde station to South Jiangsu, was located in Shanghai. Hence the data from the four radiosondes may be used to analyze the sounding situation of four regions in Jiangsu. (3) daily reanalysis data with spatial resolution of 2.5° × 2.5° from National Centers for Environmental Prediction (NCEP); (4) hourly $PM_{2.5}$ mass concentration data provided by Jiangsu Environmental Monitoring Center from 2013 to 2018.

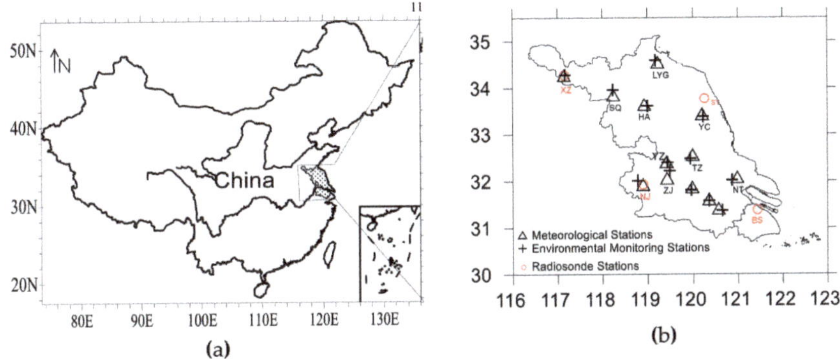

Figure 1. (**a**) The geographical location of Jiangsu province, China (UTC/GMT+08:00); (**b**) The distribution of 13 meteorological, 13 environmental monitoring and 4 radiosonde stations. (Lianyungang: in short LYG; Xuzhou: XZ; Nanjing: NJ; Nantong: NT; Suqian: SQ; Zhenjiang: ZJ; Taizhou: TZ; Huaian: HA; Yancheng: YC; Changzhou: CZ; Suzhou: SZ; Yangzhou: YZ; Baoshan: BS; Sheyang: SY).

Wind speed referred to the horizontal distance of air movement per unit time. Hourly wind speed took 1 second(s) as the time step to calculate the arithmetic average value of the 2minute(min) before the point hour. The sampling frequency of wind speed was 4 times/s.

Wind direction referred to the direction of wind. Hourly wind direction took 1 s as the time step to calculated the vector average value of 2 min before the point hour. The sampling frequency of wind direction was 1 times/s.

Relative humidity referred to the percentage of the actual partial pressure of water vapor in the air and the partial pressure of saturated water vapor under the same conditions. Hourly RHs were calculated as the arithmetic average value at 1 min before the hour. The sampling frequency of RHs was 30 times/min.

Hourly precipitation calculated the cumulative amount of 60 min before the hour. The sampling frequency of precipitation was 1 time/min.

There were four seasons in a year, namely spring (March, April, May), summer (June, July, August), autumn (September, October, November) and winter (January, February, December).

2.2. Definition of Heavy Pollution Day

The calculation method for daily average $PM_{2.5}$ concentration was based on "Ambient Air Quality Standards" (GB3095-2012) (Ministry of Environmental Protection of the PRC, 2012b). If the daily average $PM_{2.5}$ concentration above 150 µg m^{-3} was observed in no less than 3 cities of Jiangsu Province in one day, it was defined as a $PM_{2.5}$ HPDs in Jiangsu. If the daily average $PM_{2.5}$ concentration above 150 µg m^{-3} appeared in no less than three cities in a region, it was recorded as a regional $PM_{2.5}$ HPDs (RPHPDs). The pollutant concentrations of the four regions in Jiangsu Province varied greatly, so the definition of RPHPDs was very necessary.

2.3. Classification of Synoptic Patterns

We analyzed the weather situation, especially SLP, to classified synoptic patterns on RPHPDs (as well as on non-RPHPDs) by the subjective approach, the classification method was the same as Peng, et al. [21] and Huth et al. [32–34]. The results showed that, surface conditions on RPHPDs were only dominated by three patterns, respectively were equalized pressure (EQP), advancing edge of a cold front (ACF) and inverted trough of low pressure (INT) (Figure 2).

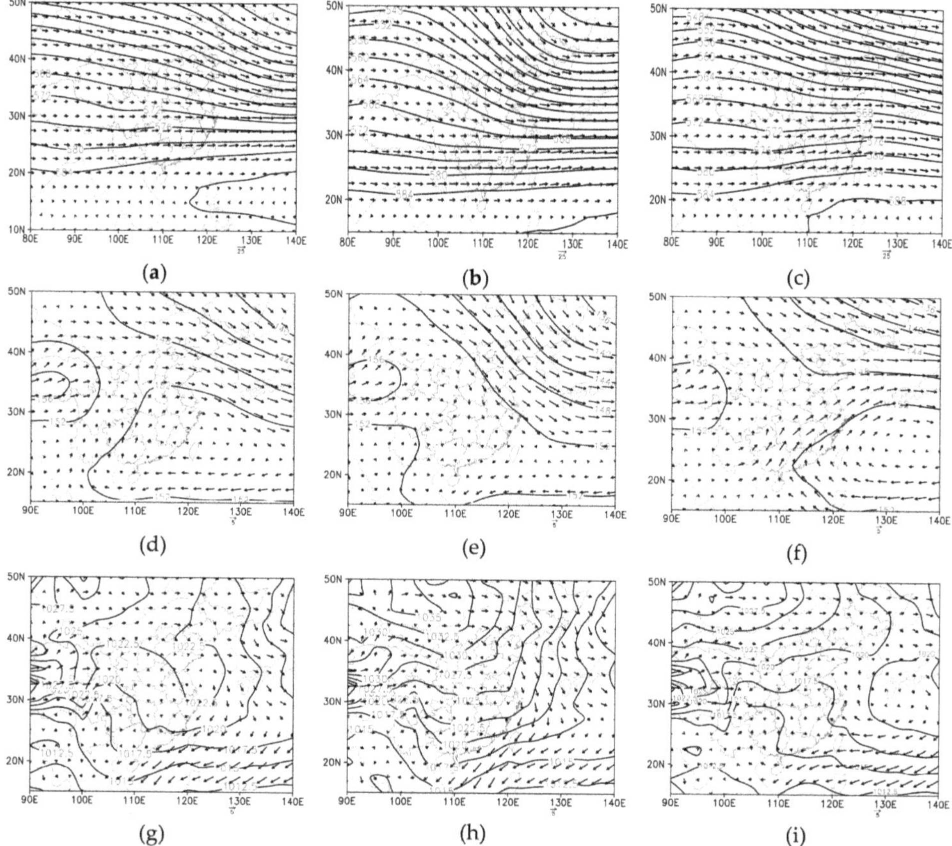

Figure 2. The synoptic situation for each type: equalized pressure (EQP: (**a,d,g**)), advancing edge of a cold front (ACF: (**b,e,h**)), inverted trough of low pressure (INT: (**c,f,i**)); 500 hPa (**a–c**), 850 hPa (**d–f**) and SLP (**g–i**). Among them, contours were pressure field (unit: hPa) and vectors were wind field (unit: m s^{-1}).

The three patterns were the necessary external condition for RPHPDs. There were any other synoptic conditions which frequently affect the flow over Jiangsu Province, but RPHPDs only occurred in the context of the three synoptic conditions. Nevertheless, non-RPHPDs were also observed under the three patterns, this was because that the synoptic pattern was the external factors causing RPHPDs [35].

(i)   EQP (Figure 2g): when the cold air was blocked in the north, the domain was controlled by equalized pressure;

(ii)   ACF (Figure 2h): when the cold air strongly advanced, the domain was controlled by the advancing edge of the cold front;

(iii) INT (Figure 2i): when the domain was controlled by the back of the weak high pressure, the high pressure receded, the inverted trough developed, and the domain was overtaken by the top of the inverted trough.

Upper and lower altitudes under three synoptic flow patterns were described as follows. At 500 hPa, the northwest westerly wind prevailed under EQP; the northwest wind prevailed under ACF; the straight west wind prevailed under INT. That is, in the meridional direction: ACF >EQP>INT (Figure 2a–c). At 850hPa, the west wind prevailed under EQP; the northwest wind prevailed under ACF; the southwest wind prevailed under INT (Figure 2d–f). In the sea-level-pressure field, Jiangsu was on the rear of the high pressure moving towards the sea under EQP; under ACF, Jiangsu was in the wedge at the front of cold high; under INT, Jiangsu was in the uniform pressure field at the front of low pressure. All the wind speeds were very low, and wind directions varied in different regions (Figure 2g–i).

Due to great differences of physical quantities in different seasons, RPHPDs were further classified (Table 1). In terms of four seasons and surface synoptic flow patterns, eight types were obtained: (a) spring INT, (b) summer EQP, (c) autumn EQP, (d) autumn ACF, (e) autumn INT, (f) winter EQP, (g) winter ACF, (h) winter INT.

Table 1. RPHPDs (regional PM$_{2.5}$ pollution days) during 2013–2017 [1]. (equalized pressure (EQP), advancing edge of a cold front (ACF) and inverted trough of low pressure (INT)).

Type	South Jiangsu	Coastal Jiangsu	Southwest Jiangsu	North Jiangsu
spring INT	\	\	2014.05.30	\
summer EQP	\	\	2014.06.07, 2014.06.15 2014.06.29	2013.06.14 2013.06.15
autumn EQP	2013.11.07	2013.11.15	2013.11.08, 2013.11.20, 2013.11.21	\
autumn ACF	\	\	\	2016.11.14
autumn INT	\	\	2013.11.09	\
winter EQP	2013.01.12, 2013.01.30, 2013.12.01, 2013.12.02, 2013.12.04, 2013.12.06, 2013.12.24, 2014.01.03, 2014.01.18, 2015.01.08, 2015.01.09, 2015.01.10, 2015.01.11, 2015.12.21, 2015.12.31	2013.12.02, 2013.12.04, 2013.12.24,2014.01.03, 2014.01.18,2014.01.30, 2014.12.29,2014.12.30, 2015.01.04,2015.01.09, 2015.01.10,2015.12.21	2013.01.28, 2013.01.29, 2013.01.30, 2013.12.01, 2013.12.02, 2013.12.06, 2013.12.24, 2014.01.02, 2014.01.03, 2014.01.18, 2014.01.30, 2015.12.31, 2017.01.03, 2017.12.31	2013.01.29, 2013.01.30, 2013.12.04, 2013.12.07, 2013.12.24,2014.01.03, 2014.01.30,2014.12.29, 2015.01.04,2015.01.10, 2015.01.26,2016.01.03, 2016.01.09,2016.12.19, 2016.12.31,2017.01.03, 2017.01.04
winter ACF	2013.01.14, 2013.01.16, 2013.12.03, 2013.12.05, 2013.12.20, 2013.12.25, 2013.12.26, 2014.01.19, 2014.01.20, 2014.02.02, 2015.02.04, 2015.02.12, 2015.02.17, 2015.12.15, 2015.12.23, 2015.12.25, 2016.01.04	2013.12.03, 2013.12.05, 2013.12.25, 2013.12.26, 2014.01.19, 2014.01.20, 2014.02.02, 2014.12.24, 2015.02.04, 2015.12.25, 2016.01.04	2013.01.13, 2013.01.24, 2013.01.26, 2013.02.23, 2013.02.24, 2013.12.03, 2013.12.04, 2013.12.05, 2013.12.15, 2013.12.20, 2013.12.25, 2013.12.26, 2014.01.19, 2015.02.12, 2015.12.15, 2016.01.04	2013.01.08,2013.02.23, 2013.12.03,2013.12.05, 2013.12.15,2013.12.20, 2013.12.25,2014.01.17, 2014.01.19,2014.02.02, 2015.02.12,2015.12.14, 2016.01.04,2016.01.10
winter INT		2013.12.08, 2014.01.31	2013.12.07, 2013.12.08, 2014.01.31, 2015.01.05, 2015.01.24	2013.12.08, 2014.01.31, 2014.02.01, 2015.01.05, 2017.12.23

[1] The date of RPHPDs,\: No RPHPDs in this type.

Considering the contingency of the data when there were less than three RPHPDs, characteristics of meteorological elements were studied only when RPHPDs were above three days in four regions and under different synoptic flow patterns (Table 1). The heavy pollution episodes which met the condition only happen in winter. The requirement was met in all four regions under winter EQP and winter ACF. While for winter INT, it was satisfied in only two regions—Southwest and North Jiangsu.

2.4. Method for Calculating MLH

The dry adiabatic curve method was used to calculate the mixing layer height (MLH). The dry adiabatic curve method was proposed by Holzworth [36,37] when studying the average maximum height of the mixing layer in some areas of the United States. As the required data were easy to obtain, the estimation method was widely used in China. The results showed that the dry adiabatic method was more practical than the methods established in the national standard GB/T 3840—91 [38] or Nozaki method [39,40] in calculating the MLH, and it was the most representative method. And its calculated results can denote the actual characteristics of the atmosphere to some extent. The daily maximum height of the mixing layer can be defined as the height below the intersection point of the dry adiabatic line and the temperature profile in the afternoon based on the daily radiosonde data at 0800 LST (Local Standard Time) and the maximum surface temperature data. This method was convenient for areas with radiosonde data and can be used for air pollution potential forecasting. Then, MLHs of Xuzhou, Sheyang, Nanjing and Baoshan radiosonde stations were calculated. Among them, Xuzhou, Sheyang and Nanjing were in North Jiangsu, Coastal Jiangsu and Southwest Jiangsu respectively; Baoshan station was a radiosonde station in Shanghai nearest to South Jiangsu. Thus, these four radiosonde stations were selected to represent the four regions (North Jiangsu, Coastal Jiangsu, Southwest Jiangsu and South Jiangsu) to analyzed the characteristics of MLH in each area.

2.5. Method for Calculating Temperature Inversion

Temperature inversion meant that the lower-layer temperature was lower than the higher-layer temperature. The lower height of a temperature inversion (LHTI) was the height (unit: m) nearest the surface when the temperature inversion layer develops. The upper height of a temperature inversion (UHTI) was the height (unit: m) at the top of the temperature inversion layer. The temperature difference in the temperature inversion was the upper temperature of the temperature inversion layer (UTTI) minus the low-level temperature of the temperature inversion layer (LTTI). The height difference in the temperature inversion (ITK) was the height of the LHTI minus the height of the UHTI. The intensity of the temperature inversion (ITI) was the temperature difference divided by the height difference multiplied by 100(unit: °C/100m): ITI = (UTTI − LTTI)/(LHTI − UHTI) * 100.

3. Results

3.1. Distribution Characteristics of RPHPDs

Statistical characteristics were obtained by the data from 2013 to 2017 and verified by the data of 2018. $PM_{2.5}$ HPDs in 13 prefecture-level cities of Jiangsu from 2013 to 2017 were shown in Figure 3. The largest number of $PM_{2.5}$ HPDs appeared in Xuzhou (109 days), followed by Huaian (83 days), Taizhou (77 days), Suqian(70 days), Nanjing (64 days), Nantong (59 days), Lianyungang (57 days), Changzhou (57 days), Wuxi (53 days), Suzhou (52 days) and Zhenjiang (51 days); the number in Yancheng was the minimum (46 days).

During 2013–2017, there were a total of 78 RPHPDs in Jiangsu (Table 1), including 33 days in 2013, 17 days in 2014, 17 days in 2015, 7 days in 2016 and 4 days in 2017, that is to say, the number of RPHPDs decreased year by year. Source emissions as a major factor in pollution declined over the years, but were not the focus of this study. This paper analyzed the role of meteorological conditions as objective factors in pollution days, so the threshold value was one of the criteria of pollution intensity. We focused on the local pollution in Jiangsu Province, the transport from other provinces were not taken into account.

According to seasons, the largest number of RPHPDs was observed in winter (65 days), followed by autumn (7 days) and summer (5 days) and the least in spring (1 days).

Considering different synoptic flow patterns, from 2013 to 2017 (Table 1), RPHPDs under EQP was the most (40 days, 51% of the cases), followed by ACF (29 days, 37%); the number under INT was the minimum (9 days, 12%).

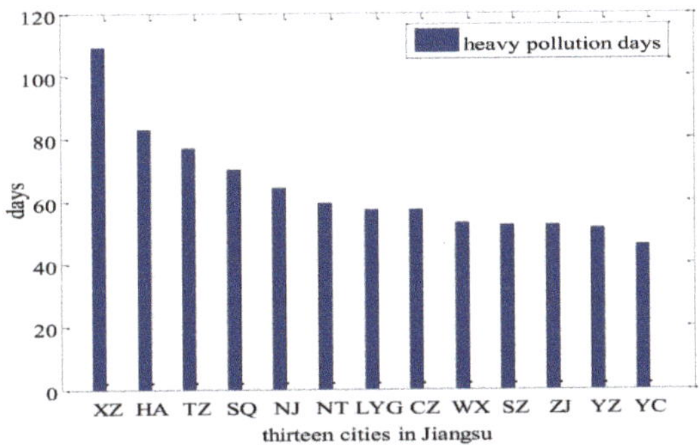

Figure 3. Heavy pollution days of 13 cities in Jiangsu Province, China (unit: days). (Lianyungang: in short LYG; Xuzhou: XZ; Nanjing: NJ; Nantong: NT; Suqian: SQ; Zhenjiang: ZJ; Taizhou: TZ; Huaian: HA; Yancheng: YC; Changzhou: CZ; Suzhou: SZ; Yangzhou: YZ; Baoshan: BS; Sheyang: SY).

3.2. Frequencies of RPHPDs under Different Synoptic Flow Patterns

The number of RPHPDs varied in different regions (Table 1). RPHPDs were 37, 26, 43 and 39 in South Jiangsu, Coastal Jiangsu, Southwest Jiangsu and North Jiangsu, respectively. The number of RPHPDs was the largest in Southwest Jiangsu, followed by North Jiangsu; the minimum number was in the Coastal Jiangsu.

The frequency of RPHPDs also varied in different seasons (Table 1). In spring, only one RPHPDs occurred in Southwest Jiangsu in May under INT. In summer, three RPHPDs and two RPHPDs respectively occurred in Southwest Jiangsu and North Jiangsu in June, which may be related to straw burning. In autumn, RPHPDs can occurred in all three patterns, but the probability was low and it only appeared in November. And the occurrence probability under EQP was higher than under ACF and INT. RPHPDs in South Jiangsu, Coastal Jiangsu and Southwest Jiangsu were one day, one day and three days respectively under EQP in autumn, while for ACF and INT in autumn, only one RPHPDs was observed in North Jiangsu and Southwest Jiangsu, respectively. RPHPDs appeared the most frequently in winter, mainly in December and January, followed by February. With regards to the three synoptic flow patterns, RPHPDs occurred more frequently under EQP and ACF and less frequently under INT. Yet no RPHPDs appeared under EQP in February, and no RPHPDs occurred in South Jiangsu under INT in winter.

The frequencies of RPHPDs under the three patterns of EQP, ACF and INT were also different (Table 1). Under EQP, RPHPDs were observed in summer, autumn and winter; under ACF, RPHPDs only occurred in autumn and winter, mainly in winter; under INT, RPHPDs occurred in spring, autumn and winter. The frequency of RPHPDs was high under patterns of g and g2 and low under INT. Moreover, the occurrence of RPHPDs was closely related to the circulation background. Under ACF, the high pressure was in the north and the cold air was active, thus RPHPDs only occurred in autumn and winter. Besides local pollution, large amounts of transportation and rapid accumulation of pollutants were also accompanied under ACF. Therefore, under EQP, the accumulation concentration of local pollutants was relatively high, and the continuous high-pressure control led to lower wind speed and stable stratification, and radiation inversions frequently happened in the night of sunny days. All these conditions finally resulted in the persistent accumulation of pollutants and a high frequency of RPHPDs. As for INT, the uniform pressure field were dominating with low wind speed and stable stratification. Located in the front of low pressure with the prevailing southeast wind, the surface convergence in Jiangsu was strong, which can easily led to the pollutant accumulation, formation of inversion weather, rise of RHs and hygroscopic growth, increasing the occurrence probability of RPHPDs.

Furthermore, frequencies of RPHPDs under eight types varied in four regions (Table 1). For South Jiangsu, the number of RPHPDs for winter ACF were the highest (17 days), followed by winter EQP (15 days), and the minimum number was for autumn EQP (1 day). In Coastal Jiangsu, the number for winter EQP (12 days) were slightly larger than that for winter ACF (11 days), followed by winter INT (2 days) and autumn EQP (1 day). In Southwest Jiangsu, the number for winter ACF (16 days) was larger than that for winter EQP (14 days), followed by winter INT (5 days), summer EQP (3 days) and autumn EQP (3 days), and the minimum number was for spring INT(1 day) and autumn INT (1 day). In North Jiangsu, the number for winter EQP (17 days) was larger than that for winter ACF (14 days), followed by winter INT (5 days) and summer EQP (2 days), and the minimum number was for autumn ACF (1 day). In all four regions, the number for winter EQP or winter ACF was significantly larger than that for other types.

Considering the contingency of the data when there were less than three RPHPDs, winter were not the only season to be considered (Table 1). Winter EQP, winter ACF and winter INT were hereinafter referred to as EQP, ACF and INT for short. Thereby, RPHPDs can be categorized into ten types: EQP in North Jiangsu (EQP_nth), EQP in South Jiangsu (EQP_sth), EQP in Southwest Jiangsu (EQP_sw), EQP in Coastal Jiangsu (EQP_cst), ACF in North Jiangsu (ACF _nth), ACF in South Jiangsu (ACF _sth), ACF in Southwest Jiangsu (ACF _sw), ACF in Coastal Jiangsu (ACF _cst), INT in North Jiangsu (INT_nth) and INT in Southwest Jiangsu (INT_sw).

3.3. Variation Characteristics of $PM_{2.5}$ and Meteorological Elements for Ten Types

3.3.1. Concentration of $PM_{2.5}$

Among the ten types, ACF_nth had the highest values in $PM_{2.5}$ concentration distribution, the high-value range of 25%–75% and the average value (Figure 4). In terms of 25%–75% high-value range and average value, for EQP, the concentration were not the highest in EQP_cst and lowest in EQP_sth; for ACF, the concentration were not the highest in ACF_nth and lowest in ACF_sth. For INT, the concentration in INT_nth was higher than that in INT_sw and the $PM_{2.5}$ concentration in ACF_sth was the lowest among the ten types. In the following section, characteristics of meteorological elements under the ten types were discussed.

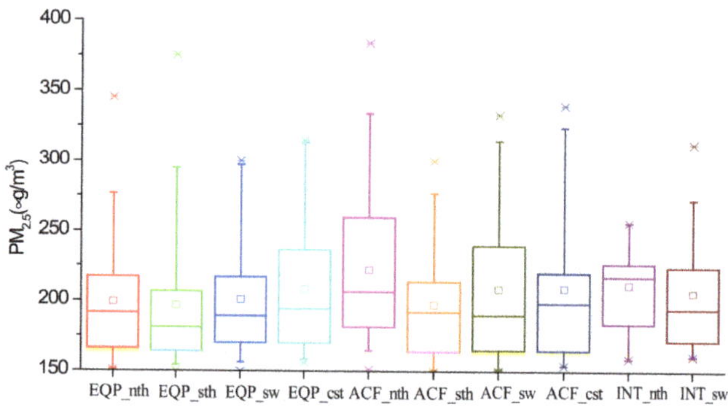

Figure 4. $PM_{2.5}$ concentration distributions of RPHPDs under the ten types. Rectangle: 25%–75% (quartile); rectangle midline: the median; end of the line: 5%–95%; ×: 1% and 99%; □: average. (equalized pressure (EQP), advancing edge of a cold front (ACF) and inverted trough of low pressure (INT).EQP in North Jiangsu (EQP_nth), EQP in South Jiangsu (EQP_sth), EQP in Southwest Jiangsu (EQP_sw), EQP in Coastal Jiangsu (EQP_cst), ACF in North Jiangsu (ACF _nth), ACF in South Jiangsu (ACF _sth), ACF in Southwest Jiangsu (ACF _sw), ACF in Coastal Jiangsu (ACF _cst), INT in North Jiangsu (INT_nth) and INT in Southwest Jiangsu (INT_sw)).

3.3.2. Wind Direction and Speed

The prevailing wind direction was different under ten different types (Figure 5). The prevailing wind in EQP_nth was the southeast direction (Figure 5a), the northwest direction in EQP_sth (Figure 5b), the southeast direction in EQP_sw and the southwest direction in EQP_cst, both followed by north direction (Figure 5c,d). In ACF_nth, the most frequent wind direction was north (Figure 5e). For ACF_sth, the direction was northwest (Figure 5f). In ACF_sw, the predominant wind direction was northwest, followed by northeast direction (Figure 5g). In ACF_cst, the wind was northwest (Figure 5h). In INT_nth, the most frequent wind direction was northwest and southeast (Figure 5i). Southeast wind prevails in INT_sw (Figure 5j). Overall, wind directions under EQP in different regions were more dispersed, whereas, under ACF and INT, they were more concentrated. Wind speed under EQP was mostly below 3 m/s. Wind speed under ACF and INT were larger, mainly distributed in 0–5 m/s and 2–5 m/s, respectively.

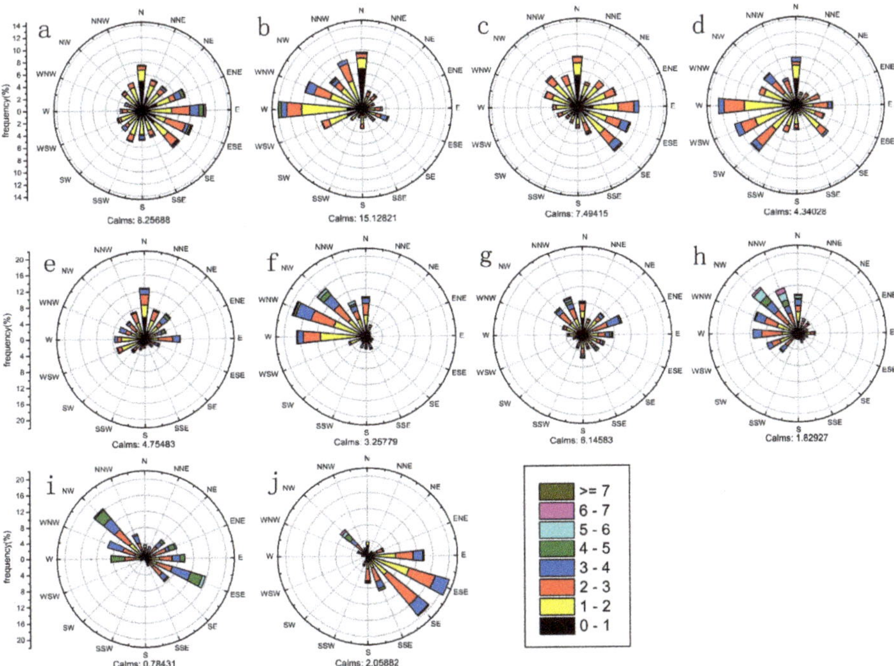

Figure 5. Wind-rose map of flow vector wind in ten types ((**a**) EQP_nth, (**b**) EQP_sth, (**c**) EQP_sw, (**d**) EQP_cst, (**e**) ACF_nth, (**f**) ACF_sth, (**g**) ACF_sw, (**h**) ACF_cst, (**i**) INT_nth, (**j**) INT_sw). The color scale represents the average wind speed during the two minutes before the integral hour point. (equalized pressure (EQP), advancing edge of a cold front (ACF) and inverted trough of low pressure (INT).EQP in North Jiangsu (EQP_nth), EQP in South Jiangsu (EQP_sth), EQP in Southwest Jiangsu (EQP_sw), EQP in Coastal Jiangsu (EQP_cst), ACF in North Jiangsu (ACF _nth), ACF in South Jiangsu (ACF _sth), ACF in Southwest Jiangsu (ACF _sw), ACF in Coastal Jiangsu (ACF _cst), INT in North Jiangsu (INT_nth) and INT in Southwest Jiangsu (INT_sw)).

3.3.3. Diurnal Variation of RHs and Wind Speed

Higher RHs was more likely to cause heavy pollution. It may be caused by the accelerated reaction and hygroscopic growth of particulate matters under high RHs conditions [41], The rise of RH altered the particle size and thus visibility. The wind was an important dynamic factor affecting the diffusion of pollutants in the boundary layer. The wind direction determined the transport direction of pollutants in the atmosphere. Wind speed determined the diffusion and dilution speed of pollutants in the atmosphere, and variations of low-level wind direction and wind speed directly affected the convergence, divergence and concentration distribution of pollutants. Therefore, RHs and wind were important meteorological elements affecting the formation and development of heavy pollution.

The average diurnal variations of RHs and wind speed were calculated for ten types. Diurnal variations of RPHPDs in different regions were like each other, but there were still some differences (Figure 6). Except for an insignificant negative correlation observed between wind speed and RHs under INT_sw (Figure 6e,f), negative correlation coefficients under the other 9 types were all in the range of 0.77–0.92, passing the 99.9% significance test.

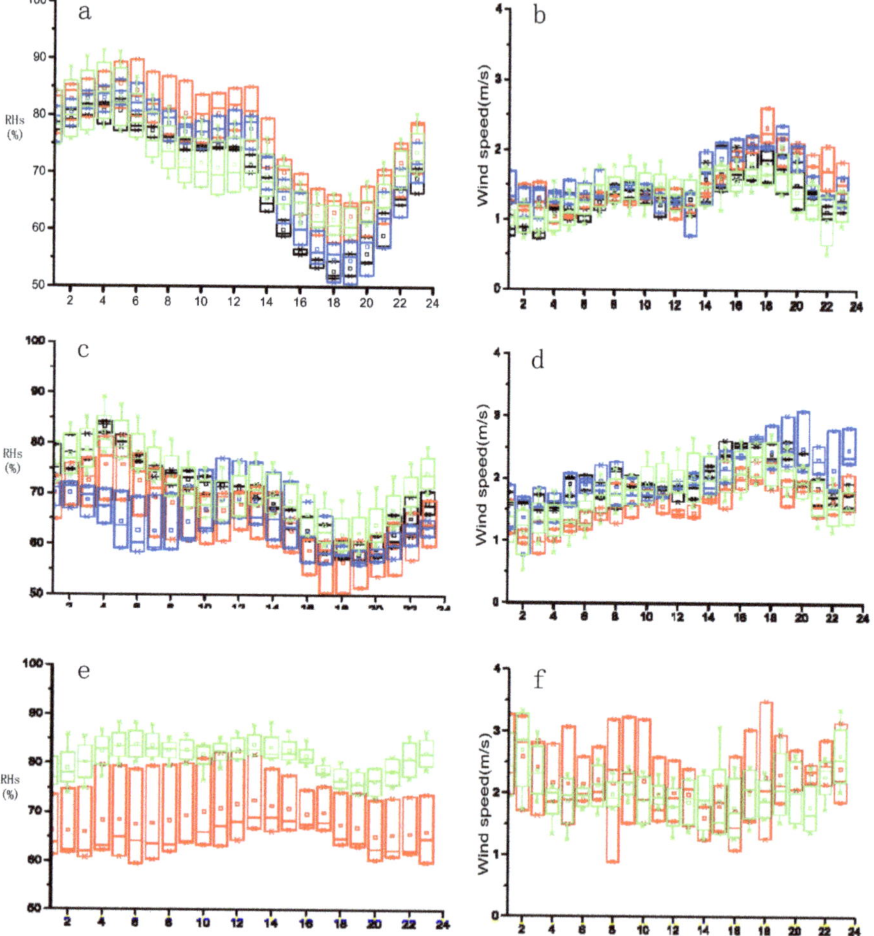

Figure 6. Diurnal variations of wind speed and RHs for EQP(**a,b**), ACF(**c,d**) and INT(**e,f**); red: North Jiangsu, black: South Jiangsu, green: Southwest Jiangsu, blue: Coastal Jiangsu, X-axis was time (unit: h); Rectangle: 25%-75% (quartile); rectangle midline: the median; end of the line: 5%-95%; ×: 1% and 99%; □: average. (equalized pressure (EQP), advancing edge of a cold front (ACF) and inverted trough of low pressure (INT)).

Diurnal variations of RHs and wind speed under EQP and ACF were more significant than INT (Figure 6a–d). The RHs decreased greatly and the wind speed increased notably in the afternoon, reaching a minimum values during 1700–1900 LST. Diurnal variation ranges of RHs and wind speed also varied in different regions. It can be seen that ranges of RHs diurnal variation were larger in Coastal Jiangsu and South Jiangsu than those in North Jiangsu and Southwest Jiangsu. Wind speeds in Coastal Jiangsu and South Jiangsu were slightly larger than that in North Jiangsu for ACF, and the range of RHs diurnal variation in Coastal Jiangsu was the smallest. The wind speed for ACF began to fluctuated and increased at 1600 LST, reaching its peak at 2000 LST and remained at a high level at night.

Compared with EQP and ACF, the diurnal variation of INT was insignificant, and diurnal variation ranges of RHs and wind speed were smaller, but wind speed was larger (Figure 6e,f). RHs in Southwest Jiangsu were significantly larger than North Jiangsu for INT.RHs decreased slightly at 1600 LST and reached the bottom value between 1900 LST and 2000 LST. Wind speed in INT_sw was within 1.8–2.7 m s^{-1} and RHs was within 76%–84% and those for INT_nth were 1.6–2.9 m s^{-1} and 65%–73%, respectively.

3.4. Threshold Values of Meteorological Elements Causing RPHPDs

Using statistical methods and meteorological condition data with RPHPDs from 2013 to 2017 to deduce the thresholds, meteorological data of 2018 were used as an independent dataset to validate the proposed thresholds. The threshold values of meteorological elements for RPHPDs were analyzed as follows.

3.4.1. Precipitation

Precipitation rarely occurred in winter on RPHPDs in Jiangsu. There were only three days, two days, one day and three days of precipitation, respectively in South Jiangsu, Coastal Jiangsu, Southwest Jiangsu and North Jiangsu on RPHPDs during 2013–2017. Probabilities of no precipitation on RPHPDs were as high as 92%, 92%, 98% and 92%, respectively. That is to say, the probability of RPHPDs without rain was above 92% with the daily and hourly precipitation below 2.1 mm and 0.8 mm. This was because the wet deposition brought by heavy precipitation can reduce the PM$_{2.5}$ concentration. When the light precipitation occurred on RPHPDs in Jiangsu, RH values increased to 83%–100%, high RH promoted the hygroscopic growth of pollutants [42], which can be explained by the Köhler's theory [43]. The rise of RH and hygroscopic growth altered the particle size and aggravated the pollutant accumulation [44–46].

3.4.2. Wind Speed and RHs

In this section, the threshold values of wind speed and RHs on RPHPDs were investigated. First, time points of RPHPDs were selected. Then average, maximum and minimum values of wind speed and RHs were calculated. Finally, the threshold values of wind speed and RHs were counted in terms of 25–75% high probability range when RPHPDs occurred. The results were shown in Figure 7. It can be seen from Figure 7a that wind speeds in ten types were very low, with an average of 1.3–2.6 m s^{-1}. Wind speeds can reached more than 5 m s^{-1} at some time points in all ten types. In terms of 25%–75% high-value range of RPHPDs, the wind speed varied within 0.5–3.6 m s^{-1}. The result was in good agreement with the conclusion of Dai et al. [47]. Specifically, the maximum wind speed was observed in INT_nth, followed by ACF_cst and INT_sw. Wind speeds in INT_nth were mainly within 1.5–3.5 m s^{-1} and the maximum value reaches 7.5 m s^{-1}. Average wind speeds in INT_nth, INT_sw and ACF_cst were all above 2 m s^{-1}.

The wind speed under EQP was minimum with the average wind speed within 1.3–1.5 m s^{-1} and the wind speed within 0.5–2.3 m s^{-1} in terms of 25%–75% high probability range. Wind speeds in different regions were basically the same. The average wind speed was 1.7–2.4 m s^{-1} under ACF, which slightly higher in South Jiangsu. The wind speed in terms of 25–75% high probability range was within 0.8–2.7 m s^{-1}. In a word, the rank of patterns based on the wind speed value on the RPHPDs was INT > ACF > EQP.

The distribution of RHs in RPHPDs was shown in Figure 7b. RHs for ten types varies greatly (21%–100%) with the average RHs within 68%–82%, the maximum was in INT_sw, followed by INT_nth; the minimum was in ACF_sth. Comparing EQP and ACF, INT had the maximum average RHs. It was found that the average RHs under EQP was higher than ACF except in Southwest Jiangsu. RHs on RPHPDs in terms of 25%–75% high probability range was within 55%–92%, with the maximum under INT (65%–92%) and RHs for EQP was basically equivalent to ACF.

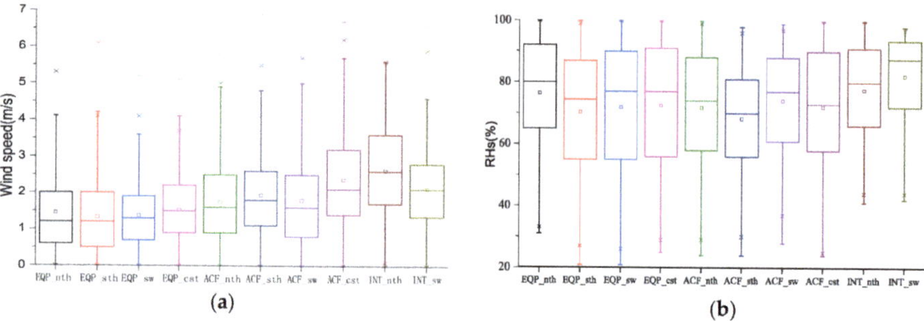

Figure 7. Wind speed (**a**) and RHs (**b**) on RPHPDs (regional PM$_{2.5}$ pollution days) under the ten types. Rectangle: 25%–75% (quartile); rectangle midline: the median; end of the line: 5%–95%; ×: 1% and 99%; □: average.

To sum up, for INT_nth and INT_sw, the light wind was more likely to cause heavy pollution than the still wind. This was because, under INT, Jiangsu was in the uniform pressure field in the front of low pressure (Figure 2i) and the southeast wind prevails (Figure 5i,j). The light southeast wind was more likely to form a weak wind field convergence near the surface than the still wind. Humidities in INT_nth and INT_sw were also higher than those in other types due to the high RHs of southeast wind. Under EQP, the entire Jiangsu Province was in the center of high pressure (Figure 2g), and wind speeds in four regions were very low. RPHPDs under ACF often occurred before the arrival of cold air moving along the middle or west path (Figure 2h). Dry cold air led to a very low RHs under ACF whose wind speed was slightly higher than that under EQP. Coastal Jiangsu were in the front of the cold front, coupled with the land-sea thermal difference, the wind speed in Coastal Jiangsu was slightly larger than that in other areas.

For the ten patterns, light wind was more likely to cause RPHPDs than still wind. The mechanism was still wind urge pollutants to accumulate in the same place, while light wind prompted pollutants to accumulate in a small area, resulting in weaker vertical shear and vertical mixing of pollutants, which will increased the scope of air pollution. Ultimately, it affected the health of more and more people. The strong wind increased the horizontal vertical shear, and the vertical mixing of pollutants was stronger, which was conducive to the diffusion of pollutants to the upper air. The explanation was consistent with the conclusion of Zhang, et al. [48].

3.4.3. ITI, ITK, LHTI and MLH

Under thermal inversion, especially low-level inversion, the vertical movement and turbulent exchange in the atmosphere were restrained, preventing the vertical transportation and horizontal diffusion of water vapor, smoke and other pollutants in the atmosphere. Thus, it led to persistent heavy pollution. The influence of inversion on the pollution diffusion was related to its ITI, ITK and LHTI [49–52].

Four radiosonde stations, Xuzhou, Sheyang, Nanjing and Baoshan, were selected as representative stations for North Jiangsu, Coastal Jiangsu, Southwest Jiangsu and South Jiangsu, respectively. The radiosonde data at 0800 LST was used to obtain the characteristics of ITI, ITK and LHTI at the lowest level in ten types of RPHPDs (Figure 8). Statistical results showed that except for type EQP_sth on January 18, 2014, inversion were observed in all RPHPDs. Temperature inversion and heavy pollution were bidirectional feedback mechanisms. A capping inversion was not conducive to the diffusion and convection of pollutants and was easy to form a static and stable weather condition that lead to accumulation of pollution. The occurrence of heavy pollution was often accompanied by temperature inversion. When pollution accumulates to a certain level, aerosol pollution can lead to temperature inversion (Zhong, 2019) [53]. ITI was 0–6.5 °C 100 m $^{-1}$, with an average of 1.6–3.1 °C 100 m $^{-1}$. ITK was within 0–651 m, with an average value 142–354 m. LHTI was within 0–1015 m, with an average value 71–407 m. When RPHPDs occurred, ITI, ITK and LHTI were respectively within 0.7–4.0 °C 100 m $^{-1}$, 42–576 m and 3–570 m in terms of 25%–75% high probability range.

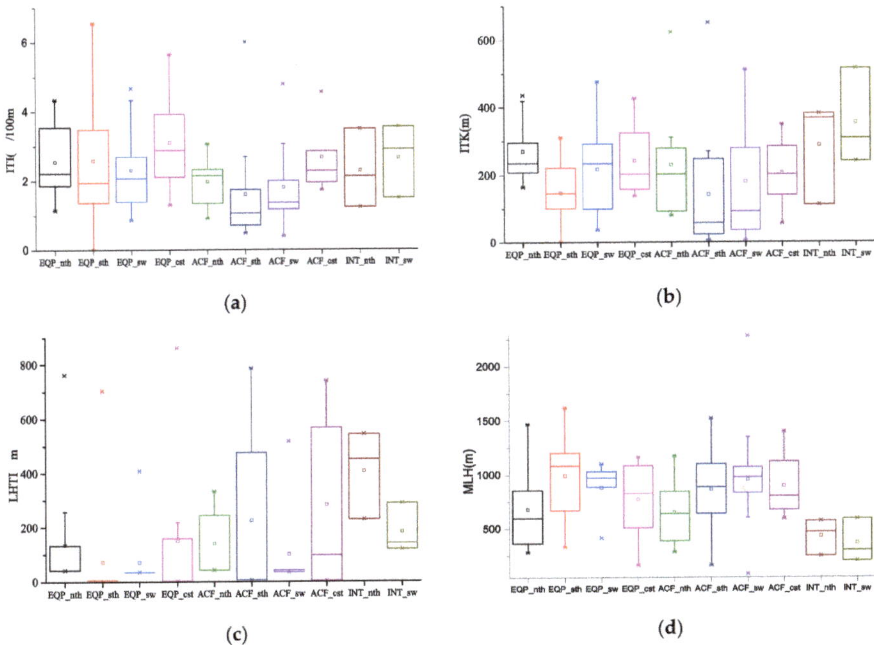

Figure 8. Inversion intensity(ITI) (**a**), height difference in the temperature inversion(ITK) (**b**), the lower height of temperature inversion (LHTI) (**c**) and mixed-layer height (MLH) (**d**) on RPHPDs (regional PM$_{2.5}$ pollution days) under the ten types. Rectangle: 25%–75% (quartile); rectangle midline: the median; end of the line: 5%–95%; ×: 1% and 99%; □: average.

Specifically, the average ITI in EQP_cst was the strongest and that in ACF_sth was the weakest (Figure 8a). Under EQP and ACF, ITI was the strongest in Coastal Jiangsu, and the weakest in Southwest Jiangsu. However, under INT, ITI in Southwest Jiangsu was slightly higher than that in North Jiangsu. For the same region, the average ITI under ACF was lower than that under EQP and INT, and the average ITK under ACF was lower than that under the other two patterns (Figure 8b). Under EQP and ACF, the largest ITK was in North Jiangsu, followed by Coastal Jiangsu, Southwest Jiangsu and South Jiangsu in turn. Under INT, ITK was larger in Southwest Jiangsu than that in North Jiangsu. Only in ACF_sth, ACF_cst and INT_nth, LHTI on a few RPHPDs was relatively higher. LHTI in other types were all below 300 m in terms of 25%–75% high probability range (Figure 8c).

The mixing layer referred to the atmospheric layer with sufficient turbulence below the discontinuous-turbulence interface [54–56]. The MLH affected the vertical diffusion of pollutants. Figure 8d showed that the MLH was relatively low (83–2284 m) when RPHPDs occurred. And the average value was less than 1 km (364–993 m). The MLH was within 200–1200 m in terms of 25%–75%

high probability range when regional RPHPDs occurred. The MLH was low and the atmospheric dispersion and dilution capability were weak, which was conducive to the accumulation of pollutants. Specifically, the mixing layer was the lowest under INT (INT_nth and INT_sw), and heights were basically the same under EQP and ACF, and MLH were in ranges of 246–740 m, 700–1200 m, 500–1100 m and 202–1100 m in terms of 25%–75% high probability range when RPHPDs occurred in North Jiangsu, South Jiangsu, Coastal Jiangsu and Southwest Jiangsu, respectively.

The ITI, ITK, LHTI and MLH were within 0.7–4.0 °C 100 m^{-1}, 42–576 m, 3–570 m, 200–1200 m, respectively. The threshold values of inversion and MLH was coincide with the results of Zhong J.T. [56], Miao Y.C. et al. [57], Huang X. et al. [58] and Ding A. J. et al. [59]. According to Ding A. J. et al. [59], the threshold values ranged extremely, the lower MLH or the greater ITI, the poorer air pollution diffusion condition. Unfavorable meteorological conditions were the necessary but not sufficient condition for RPHPDs.

3.5. Reliability Test of Threshold Values

The threshold values were deduced from the winter of 2013–2017, so the heavy pollution data in of 2018 as an independent dataset was used to conduct the reliability tests. In the winter of 2018 consists of January, February and December, with a total of 90 days, there were 29 days which meteorological conditions met both criteria: (1) the pattern was EQP, ACF or INT, (2) the eight meteorological elements were within the thresholds. There were only 11 RPHPDs that occurred in January among the 29 days, accounted for 38%. The proportion was not high, because the meteorological condition was one of factors leading to RPHPDs.

The threshold values of meteorological elements for RPHPDs in Jiangsu in winters of 2013–2017 were tested in Section 3.3. The results (Table 2) show that in the winter of 2018, RPHPDs only occurred in January, and the corresponding synoptic patterns were g and j. The frequency of RPHPDs under EQP was the highest, while only one RPHPDs under INT occurred. The number of RPHPDs in North Jiangsu was the largest, followed by Southwest Jiangsu and South Jiangsu, and no RPHPDs occurred in Coastal Jiangsu. The daily and hourly precipitation of RPHPDs in 2018 was less than 2.1 mm and 0.8 mm. The wind speed, wind direction, RHs, ITI, ITK and LHTI were all distributed within the aforementioned threshold ranges. That is to say, ITI is within 0.7–4.0 °C 100 m $^{-1}$; ITK was within 42–576 m; LHTI was within 3–570 m; MLHs in North Jiangsu, South Jiangsu and Southwest Jiangsu were also within the range.

Table 2. Dates of RPHPDs (regional PM$_{2.5}$ pollution days) and variation range of eight meteorological elements in Jiangsu in the winter of 2018. (equalized pressure (EQP), advancing edge of a cold front (ACF) and inverted trough of low pressure (INT).EQP in North Jiangsu (EQP_nth), EQP in South Jiangsu (EQP_sth), EQP in Southwest Jiangsu (EQP_sw), EQP in Coastal Jiangsu (EQP_cst), ACF in North Jiangsu (ACF _nth), ACF in South Jiangsu (ACF _sth), ACF in Southwest Jiangsu (ACF _sw), ACF in Coastal Jiangsu (ACF _cst), INT in North Jiangsu (INT_nth) and INT in Southwest Jiangsu (INT_sw)).

Weather Types	Date	Daily Precipitation (mm)	Hourly Precipitation (mm)	Wind Speed (m s^{-1})	Humidity (%)	ITI (°C 100 m^{-1})	ITK (m)	LHTI (m)	MLH (m)
EQP_nth	0116, 0117 0118, 0119 0120, 0121 0122, 0129	0.5–2.3	0–0.7	0.1–4.0	60–100	0.6–2.0	5–167	42–686	200–1188
EQP_sth	0101, 0119 0130, 0131	0–0.3	0–0.3	0.1–3.6	50–92	1.5–5.0	10–59	6–676	691–1295
EQP_sw	0119, 0120 0129, 0130	0–0.5	0–0.3	0.1–2.5	50–100	0.5–3.3	22–90	36–63	466–1109
INT_sw	0101	0	0	1.0–4.0	50–100	1.3	308	47	937

After testing, it was found that when the synoptic flow pattern were EQP, ACF or INT and eight meteorological elements were within the threshold ranges, PM$_{2.5}$ particles tend to accumulate, and air pollution diffusion conditions were poor. Unfavorable meteorological conditions were the necessary but not sufficient condition for RPHPDs. For example, in January of 2018, the probabilities of RPHPDs in South Jiangsu, North Jiangsu and Southwest Jiangsu were 2/3, 2/3 and 1/2, respectively, but RPHPDs was not necessarily to occurred as the meteorological condition was only one of factors leading to heavy pollution.

4. Conclusions

Continuous air polluting processes were jointly affected by pollutant emissions and the weather conditions that favor the accumulation of pollutants. Steady meteorological conditions were important external factors affecting air pollution. Deducing the threshold values of meteorological elements that were conducive to the pollution accumulation was very necessary to achieve a better control of air pollution. This study provided an in-depth, systematic and comprehensive research about the different synoptic situation patterns and meteorological variables on air pollution, the heavy air pollution from 2013 to 2018 in Jiangsu Province, China.

Surface conditions on RPHPDs were only dominated by three patterns, respectively were equalized pressure (EQP), advancing edge of a cold front (ACF) and inverted trough of low pressure (INT). Most RPHPDs appeared under EQP, followed by ACF, and the last days were under INT. The three patterns were the necessary condition for RPHPDs. There were any other synoptic conditions which frequently affect the flow over Jiangsu Province, but RPHPDs only occurred in the context of the three synoptic conditions. Nevertheless, non-RPHPDs were also observed under the three patterns, since the synoptic pattern was the external factors causing RPHPDs.

The threshold values of meteorological elements when RPHPDs occurred in terms of 25%–75% high probability range were as follows: The daily precipitation was below 2.1 mm; the wind speed was within 0.5–3.6 m s−1; the RHs was within 55%–92%; the thermal ITI was within 0.7–4.0 °C 100 m^{-1}; ITK was within 42–576 m; LHTI was within 3–570 m; MLH was within 200–1200 m. The MLH in terms of 25%–75% high probability range when heavy pollution occurred in South Jiangsu, Coastal Jiangsu and Southwest Jiangsu was within ranges of 246–1000 m, 163–1300 m, 162–1200 m and 200–1200 m, respectively. After validation with one-year data, threshold values were proved to be relatively reliable.

We focused on the discussion of meteorological conditions for air pollution diffusion in this paper. the source emissions as a major factor in pollution should be the main solution for air pollution control.

The influence of regional transport was not discussed. In future, if the regional $PM_{2.5}$ concentrations continue to decrease, the threshold values would remain applicable. When the threshold values were met, it represents $PM_{2.5}$ particles tend to accumulate, and air pollution diffusion conditions were poor. Unfavorable meteorological conditions always were the necessary but not sufficient condition for RPHPDs.

We took five-year data to determine threshold values, and only one-year data to finished the reliability test. RPHPDs cases of Jiangsu in 2018 were few and more cases were needed to verify the reliability of threshold values, the combined indicators need to be further constructed. In addition, this threshold needs localization correction when it extended to other regions in China or other regions around the globe. The aim was to provide an easy and quick approach to carry out heavy pollution forecasting and early warning services.

Author Contributions: Conceptualization, D.L.; validation, K.Y.; formal analysis, Z.D. and L.C.; project administration, D.L.; resources, Y.J.; writing—original draft preparation, Z.D.; writing—review and editing, D.L. All authors have read and agreed to the published version of the manuscript.

Funding: This research was jointly supported by the National Key Project of MOST (2016YFC0203303 and 2016YFC0201903), the National Natural Science Foundation of China (91544231 and 41575135), the 333 Project of Jiangsu Province (BRA2016565), the Jiangsu Meteorological Bureau Fund (KM202010), the Jiangsu Meteorological Bureau Youth Fund (KQ201913, KZ201902).

Acknowledgments: The authors would like to thank Jiansu Wei from the Meteorological Observatory of Jiangsu Province, China, Cong Li and Rong Lu from the Nanjing Meteorological Bureau of Jiangsu Province, China and Hongbin Wang from the Jiangsu Institute of Meteorological Sciences, China for their advice on text revision. Thanks to Nanjing Environmental Monitoring Center of Jiangsu Province for providing the $PM_{2.5}$ data.

Conflicts of Interest: The authors declare no conflict of interest.

References

1. Zhang, Y.; Ding, A.J.; Mao, H.T.; Nie, W.; Zhou, D.R.; Liu, L.X.; Huang, X.; Fu, C.B. Impact of synoptic weather patterns and inter-decadal climate variability on air quality in the North China Plain during 1980–2013. *Atmos. Environ.* **2016**, *124*, 119–128. [CrossRef]
2. Hampel, R.; Rückerl, R.; Yli-Tuomi, T.; Breitner, S.; Lanki, T.; Kraus, U.; Cyrys, J.; Belcredi, P.; Brüske, I.; Laitinen, T.M.; et al. Impact of personally measured pollutants on cardiac function. *Int. J. Hyg. Environ. Health* **2014**, *217*, 460–464. [CrossRef] [PubMed]
3. Tonne, C.; Halonen, J.I.; Beevers, S.D.; Dajnak, D.; Gulliver, J.; Kelly, F.J.; Wilkinson, P.; Anderson, H.R. Long-term traffic air and noise pollution in relation to mortality and hospital readmission among myocardial infarction survivors. *Int. J. Hyg. Environ. Health* **2016**, *219*, 72–78. [CrossRef] [PubMed]
4. Lee, K.; Sener, I.N. Understanding Potential exposure of bicyclists on roadways to traffic-related air pollution: Findings from El Paso, Texas, using Strava metro data. *Int. J. Environ. Res. Public Health* **2019**, *16*, 371. [CrossRef]
5. Jean, M.B.; Philippe, A.; Jérémy, G. Short-term impact of traffic-related particulate matter and noise exposure on cardiac function. *Int. J. Environ. Res. Public Health* **2020**, *17*, 1220.
6. Lim, Y.H.; Kim, H.; Kim, J.H.; Bae, S.; Park, H.Y.; Hong, Y.-C. Air pollution and symptoms of depression in elderly adults. *Environ. Health Perspect.* **2012**, *120*, 1023–1028. [CrossRef]
7. Gascon, M.; Zijlema, W.; Vert, C.; White, M.P.; Nieuwenhuijsen, M.J. Outdoor blue spaces, human health and well-being: A systematic review of quantitative studies. *Int. J. Hyg. Environ. Health* **2017**, *220*, 1207–1221. [CrossRef]
8. Zhou, C.; Li, S.; Wang, S. Examining the Impacts of Urban Form on Air Pollution in Developing Countries: A Case Study of China's Megacities. *Int. J. Environ. Res. Public Health* **2018**, *15*, 1565. [CrossRef]
9. Zhong, S.; Yu, Z.; Zhu, W. Study of the effects of air pollutants on human health based on Baidu indices of disease symptoms and air quality monitoring data in Beijing, China. *Int. J. Environ. Res. Public Health* **2019**, *16*, 1014. [CrossRef]
10. Hueglin, C.; Gehrig, R.; Baltensperger, U.; Gysel, M.; Monn, C.; Vonmont, H. Chemical characterization of $PM_{2.5}$, PM_{10} and coarse particles at urban, near city and rural sites in Switzerland. *Atmos. Environ.* **2005**, *39*, 637–651. [CrossRef]
11. Murillo, J.H.; Ramos, A.C.; García, F.A.; Jiménez, S.B.; Cárdenas, B.; Mizohata, A. Chemical composition of $PM_{2.5}$ particles in Salamanca, Guanajuato Mexico: source apportionment with receptor models. *Atmos. Environ.* **2012**, *107*, 31–41.
12. Tian, Y.Z.; Wu, J.H.; Shi, G.L.; Wu, J.Y.; Zhang, Y.F.; Zhou, L.D.; Zhang, P. Long-term variation of the levels, compositions, and sources of size-resolved particulate matter in a megacity in China. *Sci. Total Environ.* **2013**, *463–464*, 462–468. [CrossRef] [PubMed]
13. Viana, M.; Chi, X.; Maenhaut, W.; Querol, X.; Alastuey, A.; Mikuška, P.; Večeřa, Z. Organic and elemental carbon concentrations in carbonaceous aerosols during summer and winter sampling campaigns in Barcelona, Spain. *Atmos. Environ.* **2006**, *40*, 2180–2193. [CrossRef]
14. Ji, D.S.; Wang, Y.S.; Wang, L.L.; Chen, L.F.; Hu, B.; Tang, G.Q.; Xin, J.Y.; Song, T.; Wen, T.X.; Sun, Y.; et al. Analysis of heavy pollution episodes in selected cites of northern China. *Atmos. Environ.* **2012**, *50*, 338–348. [CrossRef]
15. Wang, L.L.; Nan, Z.; Liu, Z.R.; Sun, Y.; Ji, D.S.; Wang, Y.S. The Influence of climate factors, meteorological conditions, and boundary-layer structure on severe haze pollution in the Beijing-Tianjin-Hebei region during January 2013. *Adv. Meteo.* **2014**, *7*, 1–14. [CrossRef]
16. Zhang, B.; Zhao, X.M.; Zhang, J.B. Characteristics of peroxyacetyl nitrate pollution during a 2015 winter haze episode in Beijing. *Environ. Pollut.* **2019**, *244*, 379–387. [CrossRef]

17. Yu, Q.; Chen, J.; Qin, W.H.; Cheng, S.M.; Zhang, Y.P.; Ahmad, M.; Wei, O.Y. Characteristics and secondary formation of water-soluble organic acids in PM_1, $PM_{2.5}$ and PM_{10} in Beijing during haze episodes. *Sci. Total Environ.* **2019**, *669*, 175–184. [CrossRef]

18. Fu, Q.Y.; Zhuang, G.S.; Wang, J.; Xu, C.; Huang, K.; Li, J.; Hou, B.; Lu, T. Mechanism of formation of the heaviest pollution episode ever recorded in the Yangtze River Delta, China. *Atmos. Environ.* **2008**, *42*, 2023–2036. [CrossRef]

19. Liu, D.Y.; Niu, S.J.; Yang, J.; Zhao, L.J.; Lü, J.J.; Lu, C.S. Summary of a 4-year fog field study in northern Nanjing, Part 1: Fog boundary layer. *Pure Appl. Geophys.* **2012**, *169*, 809–819. [CrossRef]

20. Li, Z.H.; Liu, D.Y.; Yan, W.L.; Wang, H.B.; Zhu, C.Y.; Zhu, Y.Y.; Zu, F. Dense fog burst reinforcement over Eastern China: A review. *Atmos. Res.* **2019**, *230*, 104639. [CrossRef]

21. Peng, H.Q.; Liu, D.Y.; Zhou, B.; Su, Y.; Wu, J.M.; Shen, H.; Wei, J.S.; Cao, L. Boundary-layer characteristics of persistent regional haze events and heavy haze days in Eastern China. *Adv. Meteor.* **2016**, *11*, 1–23.

22. Wei, J.S.; Zhu, W.J.; Liu, D.Y.; Han, X. The temporal and spatial distribution of hazy days in cities of Jiangsu Province China and an analysis of its causes. *Adv. Meteor.* **2016**, *13*, 1–11. [CrossRef]

23. Chen, Z.H.; Cheng, S.Y.; Li, J.B.; Guo, X.R.; Wang, W.H.; Chen, D.S. Relationship between atmospheric pollution processes and synoptic pressure patterns in northern China. *Atmos. Environ.* **2008**, *42*, 6078–6087. [CrossRef]

24. Wu, D.; Tie, X.X.; Deng, X.J. Chemical characterizations of soluble aerosols in Southern China. *Chemosphere* **2006**, *64*, 749–757. [CrossRef] [PubMed]

25. Wu, M.F.; Wu, D.; Fan, Q.; Wang, B.M.; Li, H.W.; Fan, S.J. Observational studies of the meteorological characteristics associated with poor air quality over the Pearl River Delta in China. *Atmos. Chem. Phys.* **2013**, *13*, 10755–10766. [CrossRef]

26. Yang, L.X.; Gao, X.M.; Wang, X.F.; Nie, W.; Wang, J.; Rui, G.; Xu, P.J.; Shou, Y.P.; Zhang, Q.Z.; Wang, W.X. Impacts of firecracker burning on aerosol chemical characteristics and human health risk levels during the Chinese New Year Celebration in Jinan, China. *Sci. Total Environ.* **2014**, *476–477*, 57–64. [CrossRef]

27. Li, L.; Li, H.; Zhang, X.M.; Wang, L.; Xu, L.H.; Wang, X.Z.; Yu, Y.T.; Zhang, Y.J.; Cao, G. Pollution characteristics and health risk assessment of benzene homologues in ambient air in the northeastern urban area of Beijing, China. *J. Environ. Sci.* **2014**, *26*, 214–223. [CrossRef]

28. Ding, Y.H.; Liu, Y.J. Analysis of long-term variations of fog and haze in China in the recent 50 years and their relations with atmospheric humidity. *Sci. China Earth Sci.* **2014**, *57*, 36–46. [CrossRef]

29. Masiol, M.; Squizzato, S.; Ceccato, D.; Rampazzo, G.; Pavoni, B. Determining the influence of different atmospheric circulation patterns on PM_{10} chemical composition in a source apportionment study. *Atmos. Environ.* **2012**, *63*, 117–124. [CrossRef]

30. Kaskaoutis, D.G.; Houssos, E.E.; Goto, D.; Bartzokas, A.; Nastos, P.T.; Sinha, P.R.; Kharol, S.K.; Kosmopoulos, P.G.; Singh, R.P.; Takemura, T. Synoptic weather conditions and aerosol episodes over Indo-Gangetic Plains, India. *Clim. Dyn.* **2014**, *43*, 2313–2331. [CrossRef]

31. Zhou, L.; Xu, X.D. The correlation factors and pollution forecast model for $PM_{2.5}$ concentrations in the Beijing area. *Acta. Meteor. Sinica.* **2003**, *61*, 761–768. (In Chinese)

32. Holzworth, G.C. Estimates of mean maximum mixing depths in the contiguous United States. *Mon. Wea. Rev.* **1964**, *9*, 235–242. [CrossRef]

33. Holzworth, G.C. Mixing depths, wind speeds and air pollution potential for selected locations in the United States. *J. Appl. Meteor.* **1967**, *6*, 1039–1044. [CrossRef]

34. Xiao, Z.M.; Zhang, Y.F.; Hong, S.M.; Bi, X.H.; Li, J.; Feng, Y.C.; Wang, Y.Q. Estimation of the main factors influencing haze, based on a long-term monitoring campaign in Hangzhou, China. *Aerosol Air Quality Res.* **2011**, *11*, 873–882. [CrossRef]

35. Cheng, S.Y.; Xi, D.L.; Zhang, B.N.; Hao, R.X.; Zheng, Z.B.; Han, T.Y. Study on the determination and calculating method of atmospheric mixing layer height. *China Environ. Sci.* **1997**, *6*, 512–516.

36. Jin, M. Comparisons of boundary mixing layer depths determined by the empirical calculation and radiosonde profiles. *J. Appl. Meteor. Sci.* **2011**, *22*, 567–576.

37. Ye, D.; Wang, F.; Chen, D.R. Multi-yearly changes of atmospheric mixed layer thickness and its effect on air quality above Chongqing. *J. Meteor. Environ.* **2008**, *4*, 41–44. (In Chinese)

38. Ma, N.; Zhao, C.S.; Müller, T.; Cheng, Y.F.; Liu, P.F.; Deng, Z.Z.; Xu, W.Y.; Ran, L.; Nekat, B.; van Pinxteren, D.; et al. A new method to determine the mixing state of light absorbing carbonaceous using the measured aerosol optical properties and number size distributions. *Atmos. Chem. Phys.* **2012**, *12*, 2381–2397. [CrossRef]

39. Seinfeld, J.H.; Pandis, S.N. *Atmospheric Chemistry and Physics*; John Wiley & Sons Inc.: New York, NY, USA, 1998.

40. Petters, M.D.; Kreidenweis, S.M. A single parameter representation of hygroscopic growth and cloud condensation nucleus activity. *Atmos. Chem. Phys.* **2007**, *7*, 1961–1971. [CrossRef]

41. Su, H.; Rose, D.; Cheng, Y.F.; Gunthe, S.S.; Massling, A.; Stock, M.; Wiedensohler, A.; Andreae, M.O.; Pöschl, U. Hygroscopicity distribution concept for measurement data analysis and modeling of aerosol particle mixing state with regard to hygroscopic growth and CCN activation. *Atmos. Chem. Phys.* **2010**, *10*, 7489–7503. [CrossRef]

42. Liu, P.F.; Zhao, C.S.; Göbel, T.; Hallbauer, E.; Nowak, A.; Ran, L.; Xu, W.Y.; Deng, Z.Z.; Ma, N.; Mildenberger, K.; et al. Hygroscopic properties of aerosol particles at high relative humidity and their diurnal variations in the North China Plain. *Atmos. Chem. Phys.* **2011**, *11*, 3479–3494. [CrossRef]

43. Hennig, T.; Massling, A.; Brechtel, F.J.; Wiedensohler, A. A tandem DMA for highly temperature-stabilized hygroscopic particle growth measurements between 90% and 98% relative humidity. *J. Aerosol. Sci.* **2005**, *36*, 1210–1223. [CrossRef]

44. Dai, Z.J.; Gao, H.; Li, L.; Wang, L.J.; Wang, H.B.; Lu, X.B. Application of new data in the pollution event during the Spring Festival of 2014. *Meteor. Environ. Sci.* **2017**, *40*, 78–86. (In Chinese)

45. Liu, H.N.; Ma, W.L.; Qian, J.L.; Cai, J.Z.; Ye, X.M.; Li, J.H.; Wang, X.Y. Effect of urbanization on urban meteorology and air pollution in Hangzhou. *J. Meteor. Res.* **2015**, *29*, 950–965. [CrossRef]

46. Liu, D.Y.; Yan, W.L.; Kang, Z.M.; Liu, A.N.; Zhu, Y. Boundary-layer features and regional transport process of an extreme haze pollution event in Nanjing, China. *Atmos. Pollut. Res.* **2018**, *9*, 1088–1099. [CrossRef]

47. Liu, Y.Z.; Wang, B.; Zhu, Q.Z.; Luo, R.; Wu, C.Q.; Jia, R. Dominant synoptic patterns and their relationships with PM$_{2.5}$ pollution in winter over the Beijing-Tianjin-Hebei and Yangtze River Delta Regions. *J. Meteor. Res.* **2019**, *33*, 765–776. [CrossRef]

48. Zhang, R.H.; LI, Q.; Zhang, R.N. Meteorological conditions for the persistent severe fog and haze event over eastern China in January 2013. *Sci. China (Earth Sci.)* **2014**, *57*, 26–35.

49. Jia, M.W.; Kang, N.; Zhao, T.L. Characteristics of typical autumn and winter haze pollution episodes and their boundary layer in Nanjing. *Environ. Sci. Technol.* **2014**, *37*, 105–110. (In Chinese)

50. Pasch, A.N.; MacDonald, C.P.; Gilliam, R.C.; Knoderer, C.A.; Roberts, P. T. Meteorological characteristics associated with PM$_{2.5}$ air pollution in Cleveland, Ohio, during the 2009–2010 Cleveland multiple air pollutant studies. *Atmos. Environ.* **2011**, *45*, 7026–7035. [CrossRef]

51. Wu, D.; Liao, G.L.; Deng, X.J.; Bi, X.Y.; Tan, H.B.; Li, F.; Jiang, C.L.; Xia, D.; Fan, S.J. Transport condition of the surface layer under haze weather over the Pearl River Delta. *J. Appl. Meteoro. Sci.* **2008**, *19*, 1–9. (In Chinese)

52. Huth, R. An intercomparison of computer-assisted circulation classification methods. *Int. J. Climatol.* **1996**, *16*, 893–922. [CrossRef]

53. Huth, R. A circulation classification scheme applicable in GCM studies. *Theor. Appl. Climatol.* **2000**, *67*, 1–18. [CrossRef]

54. Huth, R.; Beck, C.; Philipp, A.; Demuzere, M.; Ustrnul, Z.; Cahynová, M.; Kyselý, J.; Tveito, O. Classifications of atmospheric circulation patterns: Recent advances and applications. *Ann. N. Y. Acad. Sci.* **2008**, *1146*, 105–152. [CrossRef] [PubMed]

55. Zhang, X.; Xu, X.; Ding, Y.; Liu, Y.; Zhang, H.; Wang, Y.; Zhong, J. The impact of meteorological changes from 2013 to 2017 on PM 2.5 mass reduction in key regions in China. *Sci. China (Earth Sci.)* **2019**, *62*, 1885–1902. (In Chinese) [CrossRef]

56. Zhong, J.; Zhang, X.; Wang, Y. Reflections on the threshold for PM$_{2.5}$ explosive growth in the cumulative stage of winter heavy aerosol pollution episodes (HPEs) in Beijing. *Tellus B* **2019**, *71*, 1–7. [CrossRef]

57. Miao, Y.; Guo, J.; Liu, S.; Liu, H.; Li, Z.; Zhang, W.; Zhai, P. Classification of summertime synoptic patterns in Beijing. *Atmos. Chem. Phys.* **2017**, *17*, 3097–3110. [CrossRef]

58. Huang, X.; Wang, Z.; Ding, A. Impact of aerosol-PBL interaction on haze pollution: Multiyear observational evidences in North China. *Geophys. Res. Lett.* **2018**, *45*, 8596–8603. [CrossRef]
59. Ding, A.; Huang, X.; Nie, W.; Sun, J.N.; Kerminen, V.-M.; Petäjä, T.; Su, H.; Cheng, Y.F.; Yang, X.-Q.; Wang, M.H.; et al. Enhanced haze pollution by black carbon in megacities in China. *Geophys. Res. Lett.* **2016**, *43*, 2873–2879. [CrossRef]

 International Journal of
Environmental Research and Public Health

Article

Combining Cluster Analysis of Air Pollution and Meteorological Data with Receptor Model Results for Ambient PM$_{2.5}$ and PM$_{10}$

Héctor Jorquera [1,2,*]  and Ana María Villalobos [3]

1   Departamento de Ingeniería Química y Bioprocesos, Pontificia Universidad Católica de Chile, Santiago 7820244, Chile
2   Centro de Desarrollo Urbano Sustentable, Pontificia Universidad Católica de Chile, Santiago 7520245, Chile
3   DICTUC S.A., Vicuña Mackenna 4860, Santiago 7820436, Chile; anamariav.i@hotmail.com
*   Correspondence: jorquera@ing.puc.cl; Tel.: +56-22-354-4421

Received: 30 September 2020; Accepted: 12 November 2020; Published: 15 November 2020

Abstract: Air pollution regulation requires knowing major sources on any given zone, setting specific controls, and assessing how health risks evolve in response to those controls. Receptor models (RM) can identify major sources: transport, industry, residential, etc. However, RM results are typically available for short term periods, and there is a paucity of RM results for developing countries. We propose to combine a cluster analysis (CA) of air pollution and meteorological measurements with a short-term RM analysis to estimate a long-term, hourly source apportionment of ambient PM$_{2.5}$ and PM$_{10}$. We have developed a proof of the concept for this proposed methodology in three case studies: a large metropolitan zone, a city with dominant residential wood burning (RWB) emissions, and a city in the middle of a desert region. We have found it feasible to identify the major sources in the CA results and obtain hourly time series of their contributions, effectively extending short-term RM results to the whole ambient monitoring period. This methodology adds value to existing ambient data. The hourly time series results would allow researchers to apportion health benefits associated with specific air pollution regulations, estimate source-specific trends, improve emission inventories, and conduct environmental justice studies, among several potential applications.

Keywords: air pollution; source apportionment; cluster analysis; health risks; residential wood burning; sustainable urban development

---

1. Introduction

Ambient air pollution is a major environmental risk worldwide. Current estimates of premature mortality brought by population exposure to ambient PM$_{2.5}$ (particulate matter with aerodynamic diameter below 2.5 μm) vary between five million [1] to ten million [2] annual deaths. This burden of disease has prompted a worldwide effort to regulate ambient PM$_{2.5}$ concentrations. A customary set of public policies targeting that overarching goal includes: ambient air pollution monitoring, emission inventories, air pollution modeling, health effects studies, and economic valuation of regulation measures (emission standards, ambient air quality standards, market-based instruments, etc.).

Ambient PM$_{2.5}$ is directly emitted by traffic, industrial, commercial, and residential sources in urban zones, and by agriculture, mining, industrial, and natural sources elsewhere. However, the resulting ambient PM$_{2.5}$ concentrations are far more complex to characterize. Some gaseous pollutants (sulfur and nitrogen oxides) undergo atmospheric oxidation, leading to sulfuric and nitric acids, which, in turn, react with ammonia to produce ammonium sulfate and nitrate particles, forming secondary PM$_{2.5}$ [3]. Furthermore, volatile (VOC) and semi-volatile (SVOC) organic compounds are

97

also oxidized in the atmosphere to become higher molecular weight compounds that condense to the particulate phase, generating secondary organic $PM_{2.5}$ [4]. The large number of organic compounds released, along with their wide range of gas/particle phase partitioning, add further complexity to ambient $PM_{2.5}$ [5]. Meteorology also plays a role in the previously mentioned chemical and physical processes, adding seasonality to ambient $PM_{2.5}$ at different time scales: wind advection moves pollutant away from their sources, temperature enhances atmospheric chemical reactions and VOC emissions, solar radiation modulates photochemical processes, relative humidity modifies fugitive PM emissions, atmospheric stability controls the degree of vertical mixing of air pollutants, rainfall brings particles and gases down to the ground, etc.

Hence, a major hurdle in quantifying population exposure to ambient $PM_{2.5}$ lies on the wide range of sources and processes that drive ambient $PM_{2.5}$ concentrations, and their variability on different time scales: diurnal, weekly, seasonal, long-term. Although ambient $PM_{2.5}$ monitoring is becoming widespread worldwide [6], a quantitative knowledge of the major sources accounting for those observed concentrations is still scarce. To obtain such knowledge, there are two customary modeling tools to conduct such a task: dispersion models and receptor models, which are described next.

Dispersion models (DM) use atmospheric emission inventories to model ambient $PM_{2.5}$ at different spatial scales: urban, regional, continental, and global [7,8]. They use atmospheric modeling to simulate how those emission sources disperse, react, deposit on the surface as they are advected by the wind field, and mixed by atmospheric turbulence. In principle, these models provide a detailed spatial and temporal representation of ambient $PM_{2.5}$ concentrations to evaluate regulatory policies at a national level. However, there are several sources of uncertainty in the emission inventories themselves and also in the parametrization of physical processes occurring in the lower atmosphere. Those two uncertainties propagate in the simulated ambient concentrations. For instance, urban zones in complex terrain pose a great challenge to current meteorological modeling tools [9], and the low-wind, stable meteorological conditions that drive air pollution episodes are difficult to be represented in current models [10]. In a review of photochemical and $PM_{2.5}$ dispersion modeling applications in the US and Canada [11], the percentage of observed variance explained by the dispersion model (r^2) varies between 0.2 and 0.6 for hourly ozone and between 0.1 and 0.6 for $PM_{2.5}$ and its major species. These are interquartile ranges and, in some cases, the model performance is better. These results represent the typical uncertainty expected for this DM approach.

Receptor models (RM) use chemical speciation of ambient $PM_{2.5}$ samples, taken on filters (daily or longer time-accumulated samples), to estimate the major sources contributing to ambient $PM_{2.5}$ mass concentrations [12–16]. They are based on the mass conservation principle applied to a set of tracer species (elements, isotopes, ions, carbonaceous aerosol, organic molecules). The required chemical speciation analyses are expensive, so most RM studies have been conducted in developed countries and have short-term results with a small proportion of long-term (multiyear) studies. A challenge for RM studies is to cover most urban zones in any given country to obtain a national level assessment of pollution sources to evaluate a regulatory policy. Regarding the uncertainty associated in RM results, one way of estimating it is to conduct a round robin exercise in which the very same data set is analyzed by different research groups. In one of these exercises conducted with a synthetic database [12], it was found that 87% of 344 RM source contribution estimates (SCE), generated by 47 participant groups, were within 50% of the correct result. Larger sources (SCE > 10% of total PM mass) were better quantified by RM, while smaller sources (SCE near 1% of total PM mass) were a challenge for most RM. In another European study (with an actual chemical speciation database), 38 RM were applied by 33 participant groups [15]. Additionally, 72% of the estimated time series of SCE were within 1-sigma uncertainty and 91% of the estimated SCE passed the z-score test. Therefore, receptor model results for SCE have a typical upper bound uncertainty of 50%.

A more recent generation of continuous ambient $PM_{2.5}$ instruments has been developed. They provide continuous information on chemical speciation [17–19] and optical properties [20–24], which also leads to source apportionment, even at a finer temporal scale (minutes, hours) than filter-based RM applications, which typically correspond to daily or weekly samples. That kind of instrumentation is more expensive than traditional ambient monitors for regulated pollutants, so it is currently less widespread than RM applications.

The preceding paragraphs show that a quantitative apportionment of major sources of ambient $PM_{2.5}$ in urban areas worldwide is yet to be accomplished, particularly in developing countries. This is the current knowledge gap.

With an ever-increasing global ambient air pollution monitoring by regulatory authorities, the possibility of extracting information out of these databases has been on the rise, estimating temporal trends in air pollution [25–29]. Furthermore, more information is expected to be collected due to the rise of low-cost ambient monitoring [30–37]. More insight into the likely sources of ambient $PM_{2.5}$ and PM_{10} can be achieved by combining meteorological and air pollution measurements. Bivariate plots of ambient concentration as a function of wind speed and direction provide information on sources contributing to ambient concentrations [38,39]. For instance, tall stack and area sources contribute most when wind speeds are higher and lower, respectively [3]. One step further consists in performing a cluster analysis (CA) of this type of bivariate representation, so that major clusters contributing to ambient concentrations may be visualized and explored [40]. One limitation of this CA approach is that there is no simple recipe to know how many clusters should be chosen in a given analysis [41]. In other words, more information is needed to decide how many clusters may be resolved in any given set of ambient air pollution data.

In a review of CA applied to air pollution analysis between 1980 and 2019 [41], it is shown that most CA applications to surface observations (i.e., air quality monitoring) consisted in exploring spatial associations among monitor sites within the same city, region, or country. The temporal evolution of those clusters was focused on their seasonal variability. In one publication [42], chemical speciation of ambient $PM_{2.5}$ was included in the cluster analysis to classify daily samples according to differences in $PM_{2.5}$ chemical composition. Then back trajectory analyses were developed to estimate air mass origins associated with each resolved cluster.

Our goal is to show that, by combining short-term RM analyses with results gotten from long-term (multiyear) CA, for the same location, it is possible to identify the major sources (clusters) in ambient air pollution data therein. With this approach, the RM results provide additional information to decide how many clusters should be considered in the CA, so both methods complement each other. We show a proof of the concept for this combined approach in three case studies of urban zones with widely different dominant sources and meteorological conditions. As far as we are aware of, this is the first time this joint analysis has been proposed for ambient air pollution data.

This combined analysis is facilitated with a set of simple rules to identify sources: $PM_{2.5}/PM_{10}$ ratios, dependence of ambient concentrations with temperature, wind speed and relative humidity, and the location of the monitoring site with respect to obvious sources: highways, industrial zones, etc. All these rules come from the literature of RM studies worldwide—see [16,43,44] for references —along with well-known results for the dispersion of stack emissions under different meteorological conditions [3] and the main features of motor vehicle emissions [45–52] as well as other combustion emissions [53–58].

The outcome of this proposed computational process is a long-term (multiyear), hourly time series of source contributions to ambient $PM_{2.5}$ and PM_{10} (and other measured air pollutants as well) suitable for different research purposes: exploring association of health effects with specific sources (example: traffic), tracking source trends, evaluate effectiveness of sector regulations, constrain sector emissions through dispersion modeling, conduct environmental justice analysis, further analysis of long-range sources using backward wind trajectories, etc. A major implication of our work is that ambient air quality databases are a rich source for further air pollution analysis worldwide, especially for source apportionment of $PM_{2.5}$ and PM_{10}.

2. Materials and Methods

2.1. Computational Methodology

We use the open access R software environment, including the openair package for air pollution analysis [59]. Openair has several dedicated functions for air pollution analysis, including the polarCluster function that groups ambient pollutant concentrations using bivariate plots. Briefly, polarCluster performs a k-means cluster analysis on the vectors [u, v, c] where u, v are the zonal and meridional wind components (sometimes referred to as the east and north wind components) and c as the pollutant concentration under analysis ($PM_{2.5}$, PM_{10}, etc.). Since all these variables have different scales, they are all standardized before the k-means algorithm proceeds [40]. This k-means clustering algorithm is well-known and has been extensively used in air pollution analyses [41], as mentioned above.

Regarding the receptor model methodology, we use already published RM results for two Chilean cities: Santiago and Temuco [60,61]. For Santiago, USEPA Positive Matrix Factorization (PMF, version 5.0) was the RM used [62], while for Temuco, the USEPA Chemical Mass Balance (Version 8.2) was the RM used [63]. Both models solve the following mass balance equation for p sources [13].

$$X_{ij} = \sum_{k=1}^{p} g_{ik}f_{kj} + e_{ij} \tag{1}$$

where X_{ij}, g_{ik}, and f_{kj} are matrices whose entries are the j-th species mass concentration measured in the i-th sample, the mass concentration from the k-th source contributing to the i-th sample, and the j-th species mass fraction in the k-th source, respectively. p is the total number of resolved sources. The residuals e_{ij} are assumed random and normally distributed.

The Chemical Mass Balance (CMB8.2) model solves Equation (1) for the case when the source profiles $\{f_{kj}\}$ have been experimentally measured for the k-th source. Hence, Equation (1) is solved for $\{g_{ik}\}$ using an effective variance least squares approach [64]. The Chemical Mass Balance model assumes that all major sources in the study area have been identified and included in its input. This model is usually run for different combinations of sources until a satisfactory solution is achieved in terms of the percentage of observed variance explained by the model.

The Positive Matrix Factorization (PMF5) model solves Equation (1) without making any assumption regarding source profiles $\{f_{kj}\}$. However, in such a case, there are more unknowns than equations in Equation (1). This means additional information must be supplied into the model [65]. Usually, source compositions $\{f_{kj}\}$ and contributions $\{g_{ik}\}$ are required to be non-negative. This is the case of software PMF5, which minimizes the following function.

$$Q = \sum_{i=1}^{n} \sum_{j=1}^{m} \left[\left(X_{ij} - \sum_{k=1}^{p} g_{ik}f_{kj} \right) / \sigma_{ij} \right]^2 \tag{2}$$

where σ_{ij} is the estimated uncertainty in the j-th species at i-th PM sample. In this approach, all observations are individually weighted by their respective uncertainties. The above minimization is carried out including the previously mentioned non-negative constrains upon compositions $\{f_{kj}\}$ and contributions $\{g_{ik}\}$.

More details on the information used for each RM application (sampling period, chemical analyses, etc.) are provided in the respective references above [60,61].

2.2. Case Studies

We have selected, as case studies to test the proposed methodology, three urban zones in Chile with widely different meteorological conditions and dominant sources in which RM results are known. Calama (22°27′S, 68°55′W, population 2017: 158,600 [66], elevation: 2500 m) is a city located within the Atacama Desert (Koppen-Geiger classification BWk [67]) and the city is close to a large mining district. The major pollution problem is ambient PM_{10}, given the desert conditions and rather strong winds that promote road dust resuspension by traffic and aeolian dust blown from the desert environment. Santiago (33°28′S, 70°40′W, population 2017: 6.85 million [66], elevation: 600 m) is the capital city, and extends along a valley surrounded everywhere by ranges and the Andes cordillera to the east. The climate is a warm temperate with dry summer (Koppen-Geiger classification CSb). Air quality regulations in Santiago were the first to be enacted in Chile and they have been successful in curbing down ambient $PM_{2.5}$ [68–71]. Nonetheless, ambient $PM_{2.5}$ in Santiago currently exceeds World Health Organization (WHO) guidelines and Chilean ambient air quality standards (AAQS), and major $PM_{2.5}$ sources are traffic, industry, commercial, and residential sources. The third case study is the city of Temuco (38°44′S, 72°35′W, population 2017: 308,600 [66], elevation: 350 m) with a warm temperate fully humid climate (Koppen-Geiger classification Cfb). In this city, residential wood burning emissions rise in colder months, leading to severe ambient $PM_{2.5}$ concentrations [60] that increase indoor concentrations as well [72].

Ambient air pollution (PM_{10}, $PM_{2.5}$, CO, SO_2, NOx) and surface meteorological measurements were downloaded from Chile's Air Quality Information System (https://sinca.mma.gob.cl/). Table 1 shows the summary of information used to conduct the analyses, where one monitoring station per city was chosen. Figure 1 shows the locations of the three monitoring sites used in our analyses. Data were screened for obvious outliers and these were removed before the computational analyses.

Table 1. Description of ambient monitoring sites analyzed.

City	Monitor	URL for Data Access	Period Analyzed
Calama	Centro	https://sinca.mma.gob.cl/index.php/estacion/index/key/236	10/2012–08/2020
Temuco	LE	https://sinca.mma.gob.cl/index.php/estacion/index/key/901	01/2009–08/2020
Santiago	LAC	https://sinca.mma.gob.cl/index.php/estacion/index/key/D13	01/2004–06/2012 [1]

[1] For the Santiago, Las Condes (LAC) site, meteorology is only available from 01/2004 through 06/2012.

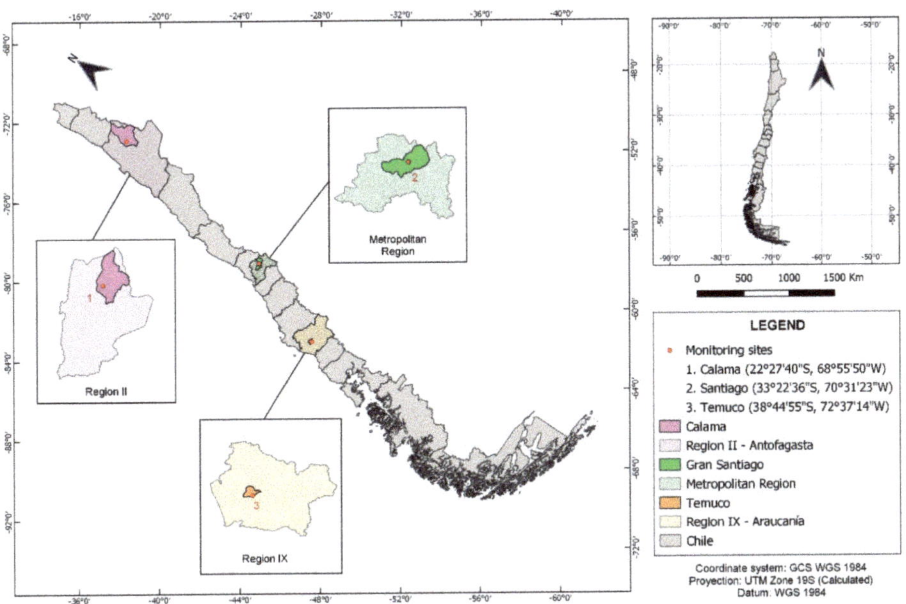

Figure 1. Geographical locations of the three cities analyzed.

2.3. Simple Comparison Rules

To map the outcome of the CA results with the RM results for the same site, we propose to use a set of rules. Some of these rules have been traditionally used in identifying RM results, such as looking at the weekly seasonality of source contributions to identify traffic or industrial sources [61,70,71,73]. The former is lower over weekends while the latter is not. Other rules use the hourly resolution of CA results to generate scatter plots and time-variability plots of ambient $PM_{2.5}$, PM_{10}, and gases (CO, NOx, SO_2), stratified by cluster, to check for specific patterns. For instance, large industrial stacks contribute more to ambient concentrations when wind speeds are higher [40] and ship emissions contribute more when ambient temperature rises [74]. A key result from scatter (X-Y) plots of air pollutants is the graphical interpretation of source compositions as 'limiting edge lines' [13,75], which define an upper (lower) edge line with the highest (lowest) Y/X ratio. Furthermore, the ratio $PM_{2.5}/PM_{10}$ is different for combustion particles than for mechanically-generated particles [3]. The proposed rules and their rationale are detailed next.

1. Residential wood burning (RWB) sources exhibit a high $PM_{2.5}/PM_{10}$ ratio, typically 0.7 or higher, with features of a single source in a $PM_{2.5}$–PM_{10} scatter plot, that is, most points lie along a straight line. RWB contributions show the highest seasonality off all resolved clusters (sources) peaking on colder months. This is a consequence of RWB emissions being driven by increasing space heating demand, so they increase in colder months whereas traffic and industrial sources remain constant all year long. Likewise, hourly RWB contributions increase when ambient temperature decreases. With respect to relative humidity (RH), RWB contributions tend to increase at higher RH, while other area sources (like fugitive dust) decrease as RH increases. RWB contributions tend to peak near midnight in colder months, unlike traffic sources that peak earlier in the evening. On a weekly basis, RWB contributions decrease less over weekends than traffic contributions do.

2. Traffic sources display a $PM_{2.5}$–PM_{10} scatter plot with a cloud of points bounded by 'limiting edge lines': an upper edge line close to a 1:1 line, characteristic of exhaust emissions, and a lower edge with a small $PM_{2.5}/PM_{10}$ ratio, typical of non-exhaust traffic emissions (i.e., road dust [76]). This is a typical signature whenever a pair of sources contribute to ambient concentrations [13,75]. Traffic contributions universally decrease on weekends, unlike industrial (or mining) sources, which are steady all year long. On a diurnal basis, traffic sources show distinctive morning and evening rush hour peaks. When plotted against wind speed, traffic sources display a negative correlation, explained by the better ventilation conditions brought by higher wind speeds. In contrast, industrial sources do not show a clear correlation or sometimes display a positive correlation because higher wind speeds brought high stack emissions down to the ground in unstable atmospheric conditions.

3. Industrial sources appear as hot spots in the CA outcome, associated with specific wind directions and high wind speeds, when tall stack contributions are relevant through fumigation processes [40]. An inspection of CA results for sulfur dioxide (SO_2) helps to clarify the location and contributions of those industrial spots to ambient $PM_{2.5}$. Since, under stable atmospheric conditions (and lower temperature and wind speeds), stack plumes will rise, their contributions to ambient SO_2 will be negligible and, therefore, traffic contributions will dominate under these circumstances. Conversely, under unstable atmospheric conditions, higher wind speed and temperatures will promote contributions from stacks, which will be the dominant ones for SO_2. In coastal areas, this mechanism will work the same way for SO_2 from ship emissions [74].

4. Aeolian dust sources appear only at high wind speeds, and, to resolve them in the CA, the number of clusters needs to be increased until those sources emerge, provided we know they are at play in the study zone.

5. As the number of clusters increases in the CA, more intermittent sources are likely to show up. Most of these intermittent sources contribute little to no ambient $PM_{2.5}$ on a long-term basis. However, they are unlikely to show up in the RM results because RM have difficulties

in resolving intermittent sources with low contributions to ambient $PM_{2.5}$ [12]. Nonetheless, they might indicate long-range, regional sources arriving to the monitoring site. Thus, these may be analyzed on their own using backward wind trajectories [77,78] to confirm their identity (natural or anthropogenic).

6. Ubiquitous area sources (like traffic) would be split by wind direction sectors as the number of clusters increases. They will all have the features presented in rule 2 above.

7. Additional gaseous measurements will provide further insights into sources' identities, comparing how those concentrations distribute across clusters. In the case of nitrogen oxides (NO, NO_2, NOx = NO + NO_2), a cluster consisting of cleaner air masses (coming from the ocean, for instance) will have low NO_2 and NOx values, while aged, long-range anthropogenic regional sources in another cluster will display larger NO_2/NOx ratios. In the case of carbon monoxide (CO), this pollutant should be apportioned mostly to traffic sources, and, hence, it would belong to traffic-related clusters. The exceptions are zones when RWB or some industrial sources are relevant. These will apportion CO as well, particularly in colder months (RWB) or under high wind speeds (industrial stack sources). SO_2 is a good tracer for large industrial sources such as copper smelters or coal-fired thermal power generation units. They may also be used as a tracer of ship emissions.

3. Results

3.1. Results for Calama

In this city in an arid environment, annual ambient $PM_{2.5}$ concentrations are below 10 $\mu g/m^3$, that is, ambient $PM_{2.5}$ satisfies current WHO guidelines [79]. Thus, we focus on ambient PM_{10} as the pollutant of concern, considering that fugitive sources, such as those windblown from desert surroundings and road dust, should be relevant. Both sources are also hard to estimate, given the intermittence of physical processes that lead to high wind speeds (gustiness conditions) in the former source, and the heterogenous processes responsible for surface dust loading on the city's streets in the latter source [51].

For this city, no RM results have been published. Nonetheless, available urban emission inventories for Calama [80] estimate that road dust (or traffic non-exhaust emission) is the dominant PM_{10} emission source, being 88% of urban emissions (Supplementary Table S1). Therefore, traffic emissions dominate PM_{10} emissions. However, no estimation of windblown dust is available for this study zone. We conduct the analysis for the monitoring station 'Centro' in Calama (see Table 1 for further details).

Figure 2 shows the outcome of the polarCluster routine when applied to ambient PM_{10} data gathered from October 2012 through August 2020. The clusters associated with windblown dust emerge with high wind speeds and wind directions between 270° and 360°, and stay the same for solutions of CA with 5–9 clusters. The other clusters include low wind speed conditions and they split into several wind direction sectors as the number of clusters increases. For simplicity, we choose a solution with five clusters and explore the results using the simple rules presented in Section 2.3 to analyze the results.

Figure 3 shows the source contributions associated with the five-cluster solution for PM_{10}. Clusters 1 and 2 are the dominant ones, followed by clusters 4, 3, and 5. The first analysis is performed looking at the time variability of this five-cluster solution. Figure S1 shows that clusters 3 and 5 present higher contributions between noon and 6 PM when wind speeds are higher in the region. Figure S2 shows scatter plots of ambient PM_{10} versus wind speed. PM_{10} concentrations from clusters 3 and 5 clearly increase with wind speed, which is distinctive of Aeolian emissions. By contrast, other clusters' contributions have little increase or decrease with wind speed (see R^2 values for instance). Cluster 3 comes from W and WNW directions and cluster 5 from NNW winds. Cluster 3 is characteristic of upslope, anabatic winds that develop all year long given the strong solar irradiation and extreme soil dryness [81]. Cluster 5 is characteristic of mountain synoptic conditions that happen in the austral fall and winter seasons, when the Pacific subtropical high weakens, favoring a higher frequency of north winds. Nonetheless, both windborne dust sources are minor contributors to a long-term average PM_{10}, as can be seen in Figure 3.

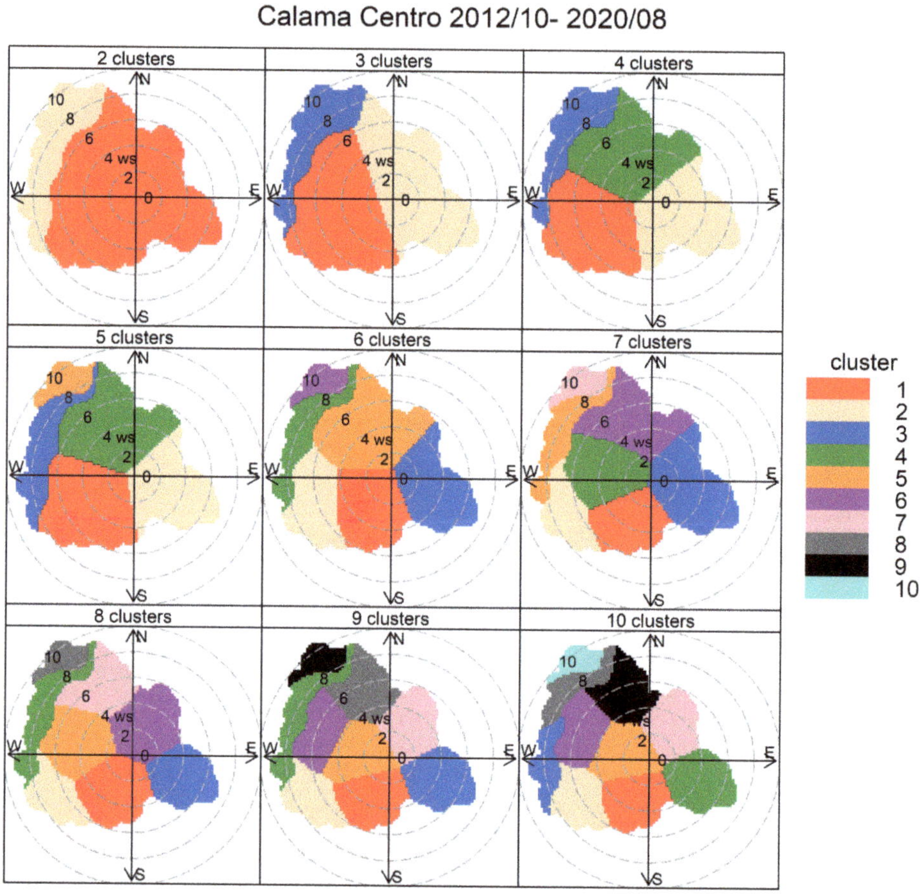

Calama Centro 2012/10- 2020/08

2 clusters | 3 clusters | 4 clusters

5 clusters | 6 clusters | 7 clusters

8 clusters | 9 clusters | 10 clusters

cluster: 1 2 3 4 5 6 7 8 9 10

Figure 2. Cluster analysis source apportionment results for ambient PM_{10} at Calama, 2012/10–2020/08, for 2–10 clusters solutions. The numbers stand for wind speed levels (m/s).

Figure 4 shows the PM_{10} time variability for clusters 1, 2, and 4. It is clear from this figure that these three clusters correspond to traffic sources: their ambient contributions have distinctive peaks during morning and evening traffic rush hours, and they significantly decrease over the weekends. The same behavior can be seen in Figure 5, where the NOx time variability is plotted for clusters 1, 2, and 4. Figure 6 shows scatter plots of $PM_{2.5}$ and PM_{10} by cluster. In clusters 1, 2, and 4, the points lie between two limiting 'edge lines': one upper edge with high $PM_{2.5}/PM_{10}$ ratio (traffic exhaust) and a lower edge with a low $PM_{2.5}/PM_{10}$ ratio (lower that 0.1), characteristic of non-exhaust traffic emissions [46,51]. Thus, any point in the plot for those three clusters can be regarded as an air mass that arrives to the monitor site with a mixture of those two traffic emissions, as it is customarily presented in the receptor modeling literature [13,75].

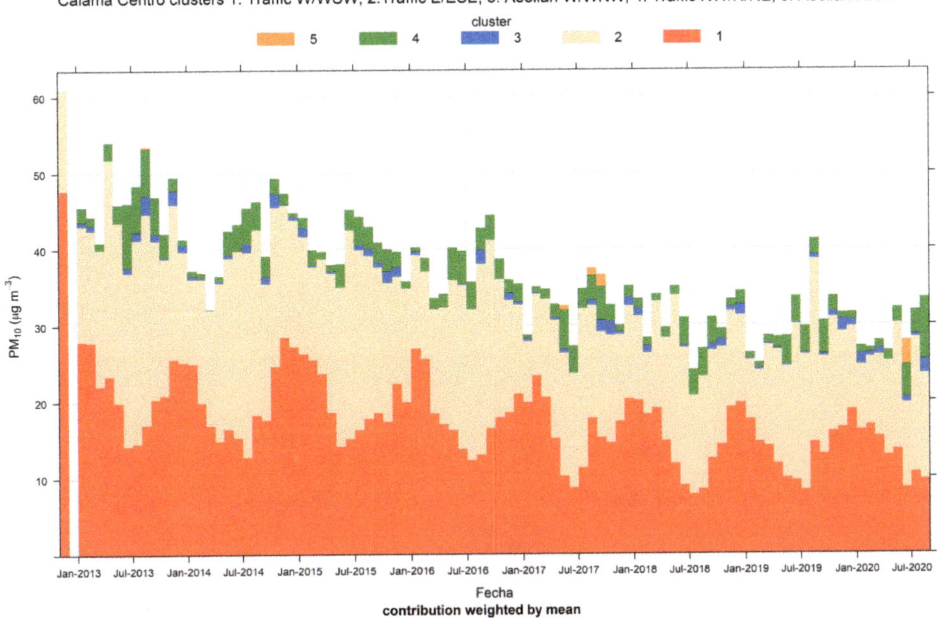

Figure 3. Cluster analysis source apportionment results for ambient PM_{10} at Calama, 2012/10–2020/08, for a 5-cluster solution.

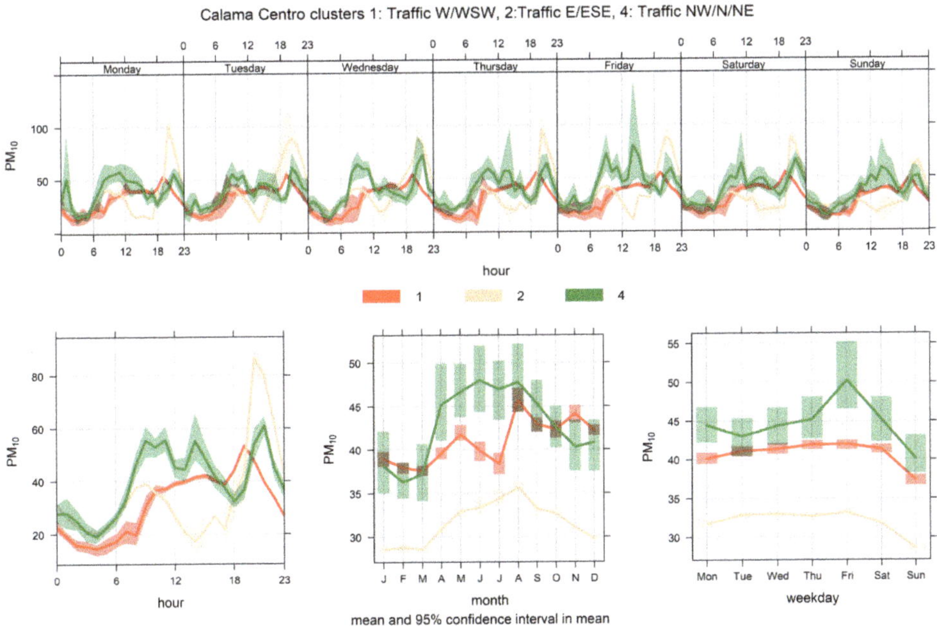

Figure 4. Time variation plot for cluster analysis source apportionment results for ambient PM_{10} at Calama, 2012/10–2020/08, for a five-cluster solution. Only clusters 1, 2, and 4 are plotted for clarity's sake.

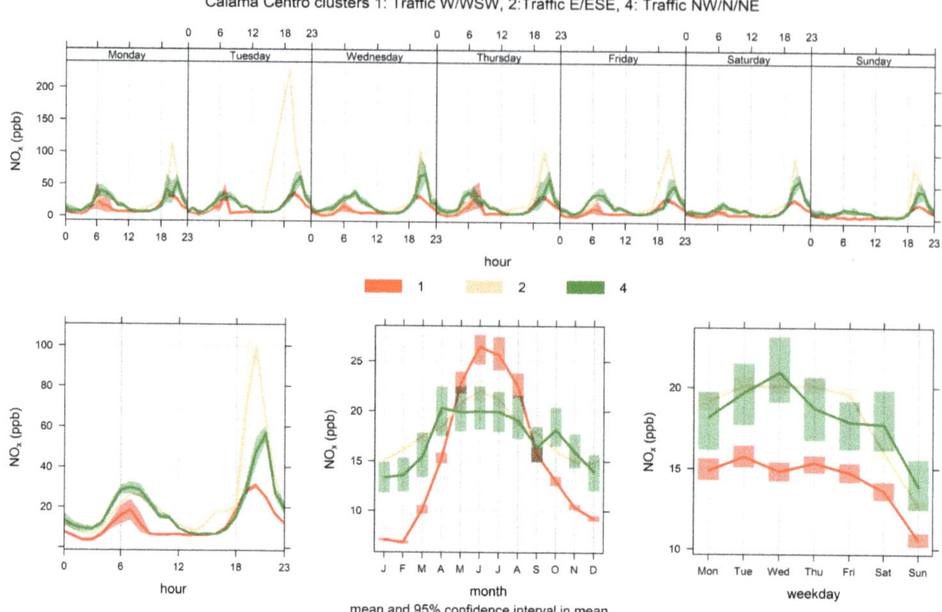

Figure 5. Time variation plot for cluster analysis source apportionment results for ambient NOx at Calama, 2012/10–2020/08, for a five-cluster solution. Only clusters 1, 2, and 4 are plotted for clarity's sake.

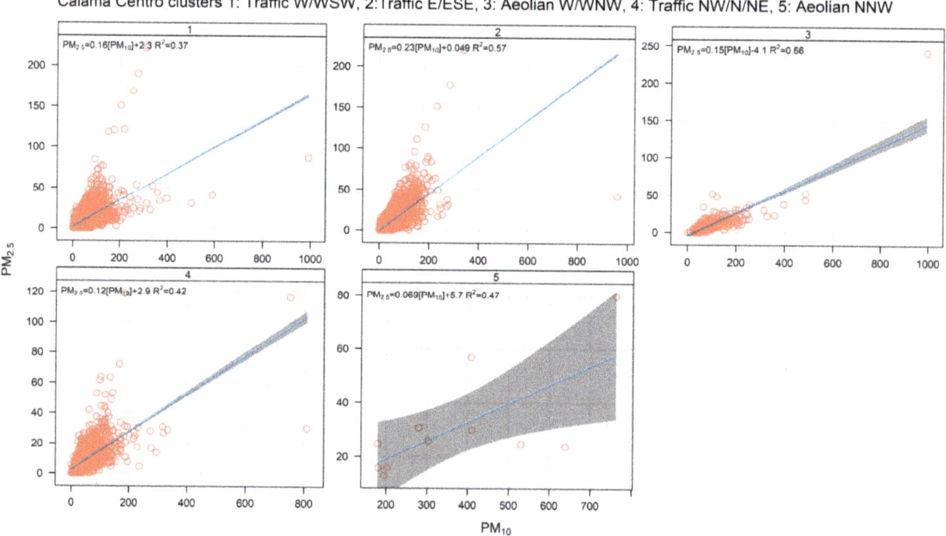

Figure 6. $PM_{2.5}$–PM_{10} scatter plot by cluster for a five-cluster solution for Calama.

3.2. Results for Temuco

In this city located in a wet temperate climate, RWB is the major source of ambient $PM_{2.5}$. RM results show that, in the winter season, RWB contributions to ambient $PM_{2.5}$ vary between 70% and 100% [60]. These RM results were found for a monitoring site located 700 m NW of the air quality monitoring site that we analyze here (denoted as LE by its Spanish name 'Las Encinas'). The data cover the period from January 2009 through August 2020. In this case study, we use temperature instead of wind speed for conducting the CA. Had we chosen the traditional CA, we would have found difficulties in interpreting the resulting clusters because of an overlapping of high $PM_{2.5}$ contributions at low wind speeds, which mixes traffic and RWB contributions (results not shown). By using temperature as an input variable, we are able to extract the RWB contribution and separate it from the traffic contribution. This approach works well precisely because the highest RWB impacts occur at nearly midnight in colder months when traffic sources are minimal and RWB emissions are the highest. Furthermore, since under those stable atmospheric conditions, wind direction is highly variable because of wind meandering effects, RWB contributions should come from all wind directions, defining a single cluster enclosing the origin of coordinates in the bivariate plot. This geometric feature helps in identifying RWB sources, even when they are not the dominant ones (see Section 3.3).

Figure 7 shows the cluster analysis results for two to eight clusters in Temuco, Las Encinas (LE) site. It can be clearly seen that a 'central cluster' develops for solutions with three or more clusters. For seven or more clusters, this central cluster splits itself into a pair of low-temperature and high-temperature clusters. They both correspond to RWB (results not shown here). For simplicity, we choose a three-cluster solution to identify the sources.

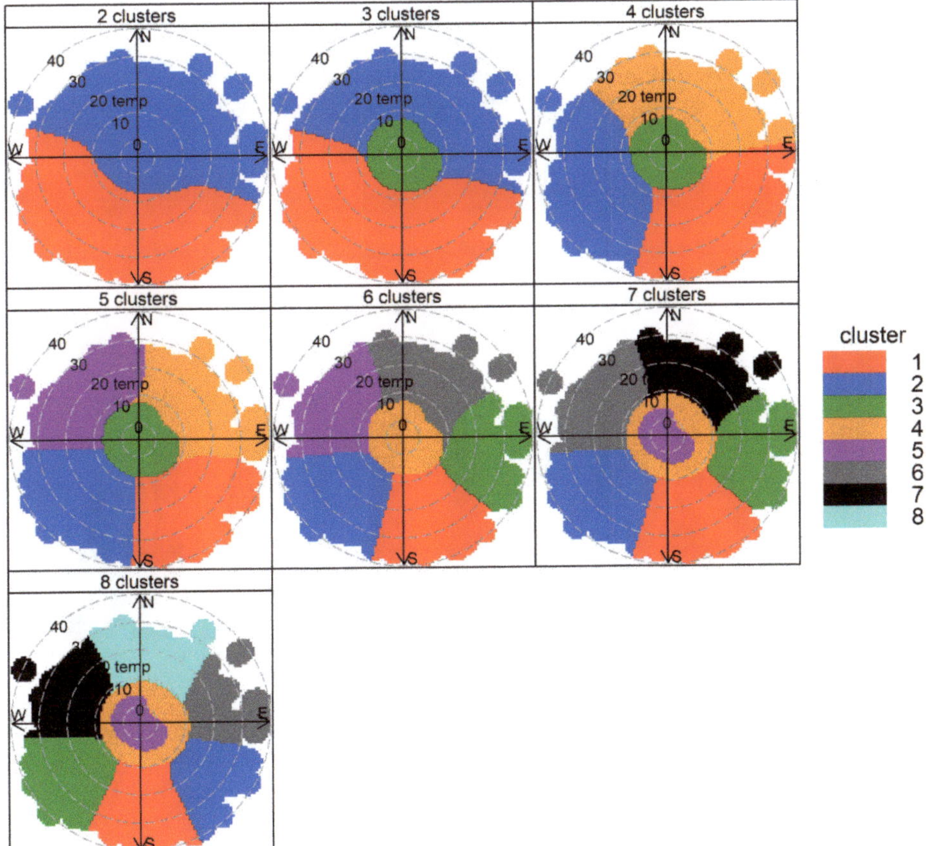

Figure 7. Cluster analysis source apportionment results for ambient $PM_{2.5}$ at Temuco, 2009/01–2020/08, for 2–8 cluster solution.

Figure 8 shows the monthly source contributions of the three-cluster solution found in Temuco. Source 3 is the dominant one, followed by sources 1 and 2. Figure 9 shows the time variability of those three clusters. Clusters 1 and 2 show lower seasonality than cluster 3, which has dominant contributions near midnight. Figure S3 shows the pollution rose by the cluster. Cluster 3 has contributions for low wind speeds and from all wind directions, while clusters 1 and 2 are each associated with a set of narrow wind directions. Figure 10 displays $PM_{2.5}$-PM_{10} scatter plots. Clusters 1 and 2 have two clear limiting edge lines—characteristic of traffic sources—while cluster 3 points lie along a straight line with a high slope near 1. Figure S4 shows CO time variability by cluster. The cluster patterns mimic the ones already shown in Figure 9 for $PM_{2.5}$. Figure 11 show scatter plots of $PM_{2.5}$ against the temperature. Cluster 3 has a different behavior and higher $PM_{2.5}$ concentrations than in clusters 1 and 2. Therefore, the CA results indicate that residential wood burning is the dominant source of ambient $PM_{2.5}$, which is followed by traffic sources (clusters 1 and 2) with lower contributions.

We now turn to a quantitative comparison of the above CA results with RM results obtained for the period July–September 2014 at a monitoring site 700 m NW of the LE site. Table 2 shows a comparison of the average sources identified with CA and those resolved using the chemical mass balance RM (CMB8.2) with organic molecular markers [60]. The molecular markers measured in ambient $PM_{2.5}$ samples were used to apportion organic carbon (OC) among different combustion sources (gasoline, diesel, wood, coal, natural gas) using Equation (1) with known source profiles $\{f_{kj}\}$. Afterward, organic source contributions to $PM_{2.5}$ mass were calculated from those source contributions to OC and specific OC/$PM_{2.5}$ mass ratios for each source [82–85].

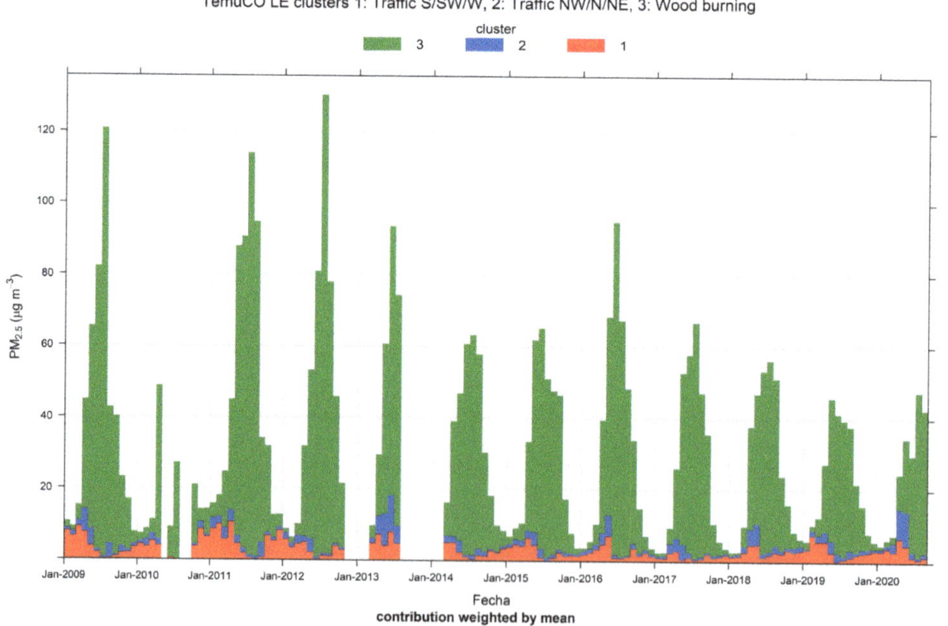

Figure 8. Cluster analysis source apportionment results for ambient $PM_{2.5}$ at Temuco, 2009/01–2020/08, for a three-cluster solution.

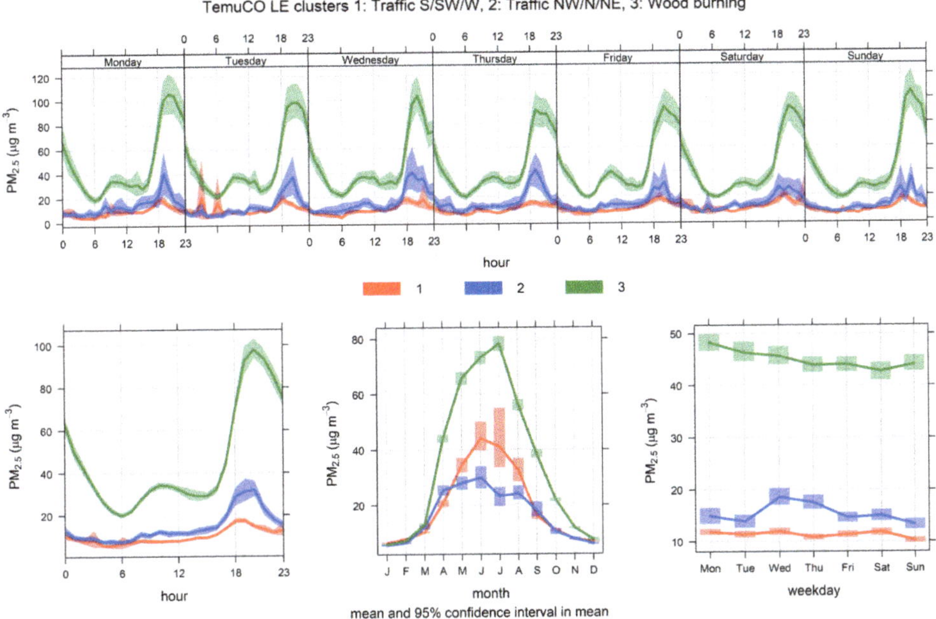

Figure 9. Time variation plot for cluster analysis source apportionment results for ambient $PM_{2.5}$ at Temuco, 2009/01–2020/08, for a three-cluster solution.

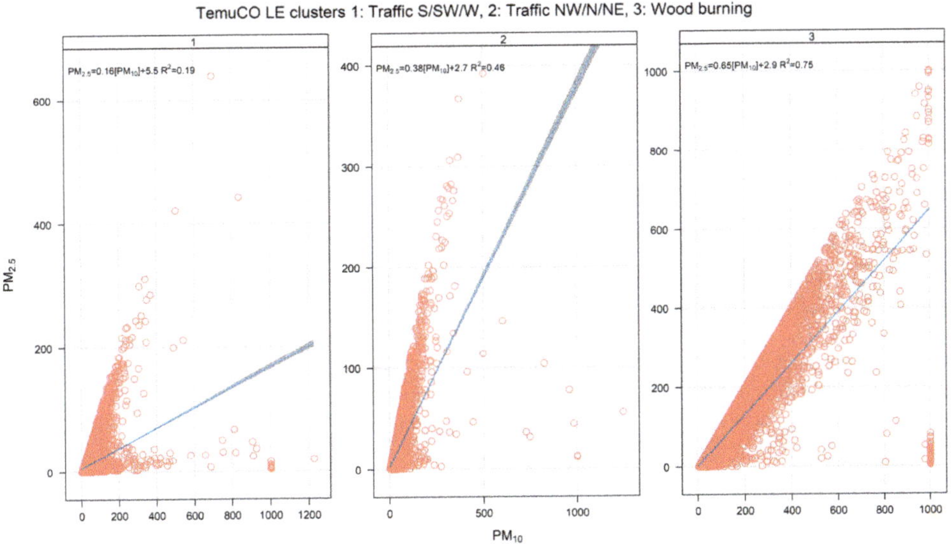

Figure 10. $PM_{2.5}$–PM_{10} scatter plot by cluster, for a three-cluster solution for Temuco.

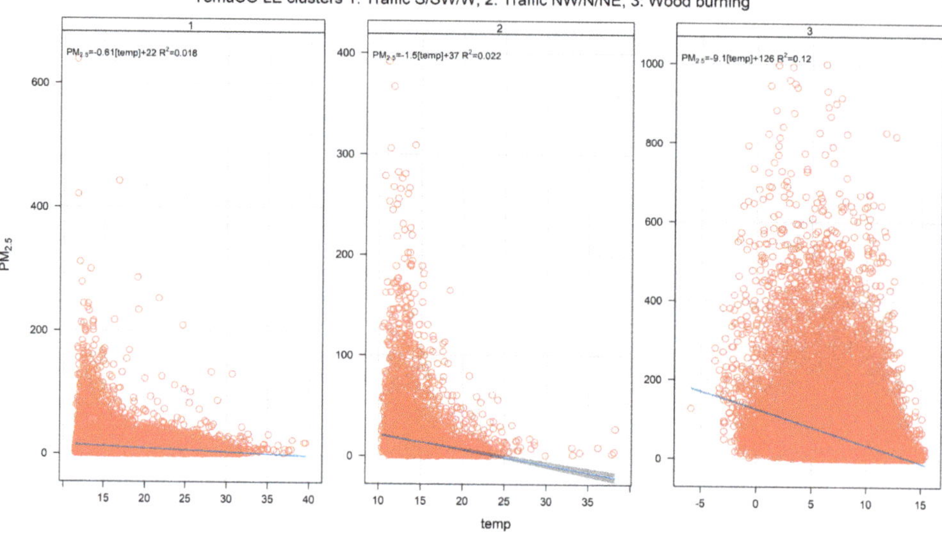

Figure 11. PM$_{2.5}$-temperature scatter plot by cluster for a three-cluster solution for Temuco.

Table 2. Comparison for cluster analysis (CA) and receptor model (RM) source apportionment estimates (±standard error) for Temuco, July–September 2014 [1] in (µg/m^3).

Period	RWB [2] (CA)	RWB (RM)	Traffic (CA)	Traffic (RM)
Week 1	35.3 ± 4.1	42.7 ± 5.9	0.9 ± 0.5	2.2 ± 0.2
Week 2	64.3 ± 7.1	75.9 ± 7.1	0.5 ± 0.5	2.4 ± 0.5
Week 3	15.2 ± 2.8	18.6 ± 5.0	3.2 ± 1.0	1.4 ± 0.1
Week 4	47.9 ± 5.6	52.6 ± 6.3	0.0 ± 0.0	2.3 ± 0.2
Week 5	52.9 ± 6.6	56.9 ± 6.7	3.7 ± 1.1	2.6 ± 0.2
Week 6	43.7 ± 4.4	42.8 ± 6.1	2.7 ± 1.0	2.5 ± 0.2
Week 7	23.1 ± 4.2	23.5 ± 5.0	5.1 ± 1.1	1.5 ± 0.1
Week 8	57.6 ± 9.0	48.6 ± 7.8	1.1 ± 0.3	2.8 ± 0.2
Average	42.5	45.2	2.2	2.2

[1] Supplementary material in Reference [60] provides further details and data. [2] Residential wood burning sources.

For the CA traffic sources, we compare them with the sum of diesel exhaust emissions, vegetative detritus (both resolved by CMB8.2), and dust (sum of Al, Si, Fe, Ca, and Ti oxides) to consider the non-exhaust contributions included in the CA (i.e., road dust). For the residential wood burning source, the comparison is more elaborated. We have found that, in Temuco, coal combustion is also used for space heating (identified using picene as organic tracer), so we have added RWB contributions (identified using levoglucosan as an organic tracer) to those from residential coal combustion since both processes happen under the very same environmental conditions. Furthermore, there is a substantial contribution of secondary organic aerosol (SOA) in Temuco, which is ascribed to the inefficient burning of wood. This combustion process is known to release semi-volatile organic compounds [53], which quickly oxidize and, thus, generate secondary organic aerosols [86]. The CMB8.2 RM cannot resolve secondary sources, so the unresolved organic carbon (OC) is denoted as 'Other OC'. This 'Other OC' is identified as SOA by its high correlation with water soluble organic carbon (WSOC), indicating a high degree of molecular oxygenation. Therefore, we added the corresponding SOA contribution to the RWB (and coal combustion contribution) to get the 'RWB RM' entry in Table 2.

We show in Table 2 the previously mentioned comparisons for the 8-week ambient monitoring campaign reported in Reference [60]. For every weekly sample, we computed the average CA contributions for both sources (Traffic: clusters 1 and 2, RWB: cluster 3), considering the same five sampling days per week as in the ambient measurement campaign. Standard errors for all estimates (from CA and RM results) were computed from error propagation. For both major sources that can be resolved with CA, the agreement is good with very similar winter average results. The zero value for one week in the traffic CA contribution is a result of trying to identify a weak signal in a data set dominated by a single source (RWB in this case). This is the usual outcome both in CA and RM analyses.

Therefore, the comparison results in Table 2 show that the proposed CA analysis is able to capture the major sources contributing to ambient $PM_{2.5}$ in this case study of a zone dominated by residential wood burning emissions. For the (smaller) traffic contributions, the methodology has difficulties in capturing the temporal variation, even though, on average, the results are similar to the RM results.

3.3. Results for Santiago

In this case study of a large metropolitan area, we focus the analysis on a monitoring site located near the east edge of the city at a higher elevation in Santiago's basin (henceforth, denoted as LAC, which is short for its Spanish name 'Las Condes'). For that site, a previous RM study [61] has shown that traffic sources and residential wood burning are the dominant ones, although regional sources also contribute to ambient $PM_{2.5}$. That study was conducted with another RM, Positive Matrix Factorization (PMF), using elemental concentrations in ambient $PM_{2.5}$ daily samples as input data.

Figure 12 shows the results of CA for 2 to 10 candidate clusters, using temperature instead of wind speed as input variable, to resolve RWB contributions. A central cluster with the lowest ambient temperatures shows up for solutions with seven or more clusters, suggesting this is the RWB source. We choose an eight-cluster solution and we present these results here. Figure 13 shows the source contributions brought by this eight-cluster solution. Clusters 3 and 4 are the ones with the largest source contributions, followed by cluster 7 (that groups the lowest ambient temperatures) and clusters 1 and 6. The rest of the clusters are of minor relevance. Figure S5 shows that cluster 3 has predominant WSW directions while cluster 4 includes ENE and E directions. Figure 14 shows the $PM_{2.5}$ time variability but only for the five major contributing clusters. It can be seen that cluster 4 has morning and evening peaks, coincident with rush hour traffic conditions. In cluster 3, its contribution increases from morning to early afternoon, indicating the arrival of traffic contributions from the city, brought by anabatic winds. This rise is followed by a decline of contributions later in the evening. This daylight behavior of anabatic winds and pollution transport toward the east side of the city has been recently measured and modeled for black carbon particles in Santiago [87]. Cluster 7 is the only one that does not decrease over weekends, unlike the other clusters shown in Figure 14, which decrease over weekends. This temporal pattern supports the identification of cluster 7 as the RWB sources.

Figure 15 shows $PM_{2.5}$-PM_{10} scatter plots by cluster, showing that, in cluster 7, the data have the highest slope of all, suggesting that this is the RWB source. For clusters 1–5 and 8, the scatter plots show that the respective data points have upper and lower limiting edge lines, as expected for traffic sources.

To better identify this eight-cluster solution, we present in Table 3 the monthly average $PM_{2.5}$ contribution by cluster, for year 2004 for which we have RM results in this same monitoring site.

Table 3. Monthly source contributions to ambient PM$_{2.5}$ by cluster for Santiago, year 2004 (μg/m^3).

Month	C1	C2	C3	C4	C5	C6	C7	C8
1	2.1	0.3	10.4	5.8	0.0	0.9	0.0	0.1
2	2.7	0.6	11.4	9.1	0.0	1.8	0.0	0.0
3	1.8	0.3	11.0	10.7	0.0	2.4	0.0	0.2
4	1.7	0.2	5.8	13.3	0.02	3.7	0.7	0.5
5	0.4	0.1	5.7	17.4	0.0	6.6	6.5	0.4
6	0.1	0.0	4.0	16.7	0.0	5.9	4.5	1.3
7	0.0	0.0	3.1	16.0	0.0	4.7	3.9	0.9
8	0.2	0.0	4.7	9.2	0.0	3.2	6.8	0.4
9	0.4	0.0	6.3	10.8	0.0	3.1	2.1	0.4
10	1.0	0.1	6.3	8.0	0.0	2.8	0.6	0.3
11	1.5	0.3	7.7	6.0	0.0	3.2	0.1	0.4
12	2.4	0.5	10.4	5.7	0.0	1.5	0.0	0.3

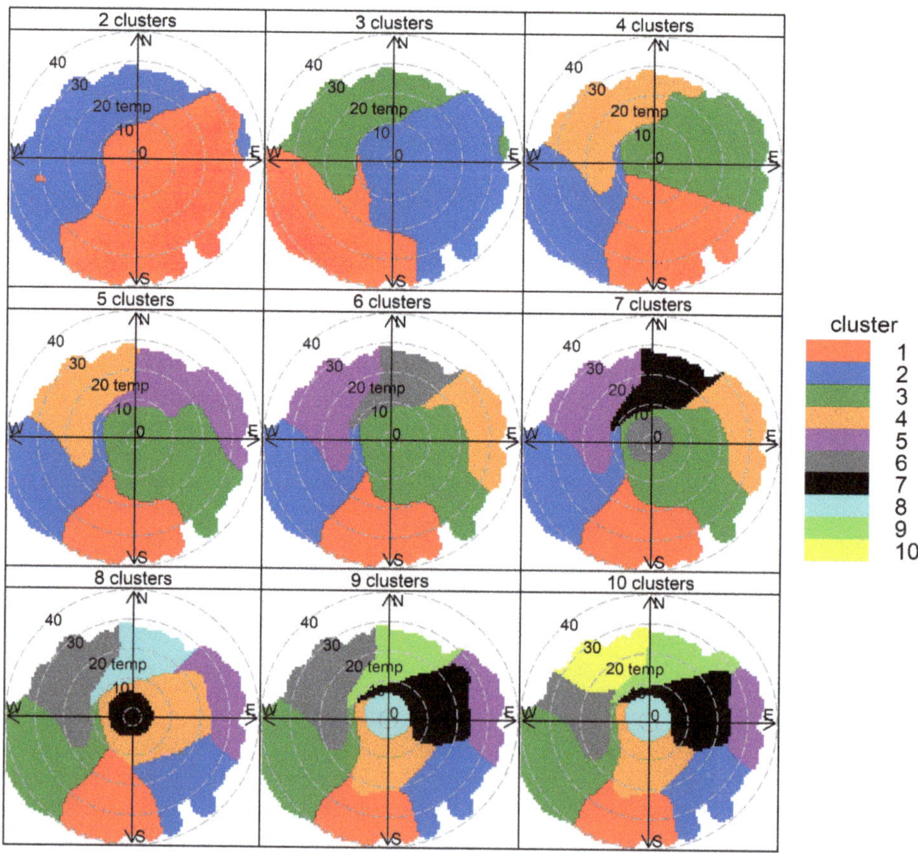

Figure 12. Cluster analysis source apportionment results for ambient PM$_{2.5}$ at Santiago, 2004/01–2012/06, for 2–10 cluster solutions.

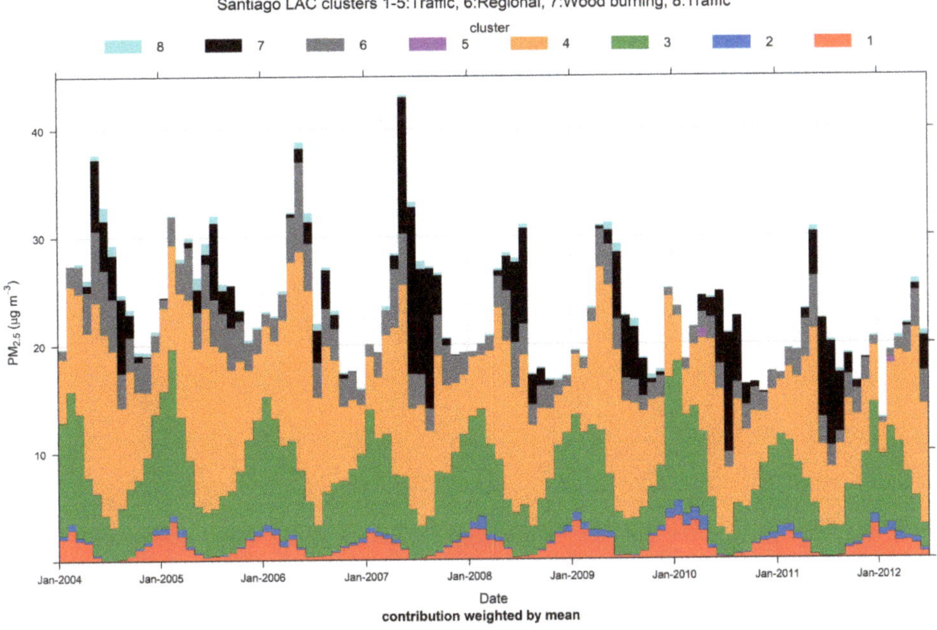

Figure 13. Cluster analysis source apportionment results for ambient PM$_{2.5}$ at Santiago, 2004/01–2012/06, for an eight-cluster solution.

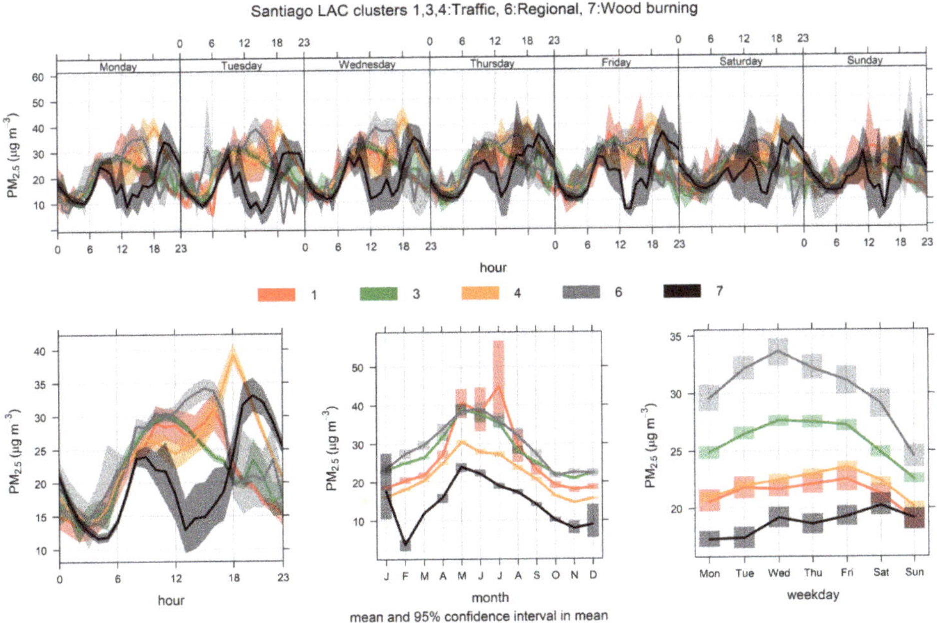

Figure 14. Time variation plot for the cluster analysis source apportionment results for ambient PM$_{2.5}$ at Santiago, 2004/01–2012/06, for an eight-cluster solution. For clarity's sake, only the five major clusters are included.

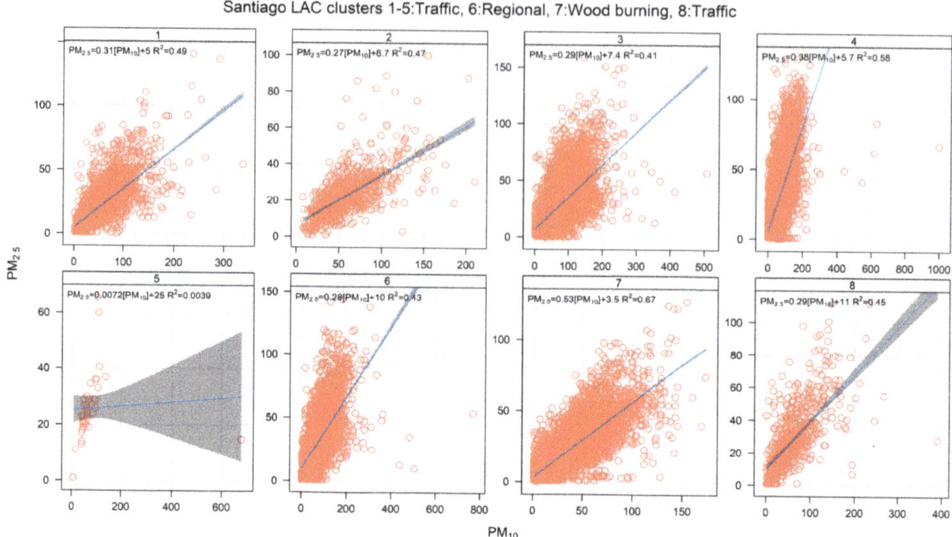

Figure 15. $PM_{2.5}$–PM_{10} scatter plot by cluster, for an eight-cluster solution for Santiago.

From Table 3, we can see that clusters 4, 7, and 8 increase their contributions during the fall and winter season, whereas clusters 1, 2, and 3 show minimum contributions in those colder seasons. The reason for this different behavior has to do with meteorological conditions. In the fall and winter seasons, subsidence conditions promote a low level of thermal inversion layers over Santiago's valley, leading to shallow planetary boundary layers (PBL) [88] and blocking transport of emissions from Santiago's lower valley (dominant wind direction for clusters 1, 2, and 3 as seen on Figure S5). This also explains why local sources (clusters 4, 7, and 8) increase their contributions during the fall and winter. This topography-induced effect has been reported before for total ambient PM concentrations [89] and, more recently, in the simulation of black carbon transport from Santiago toward the Andes mountains east [87].

Regarding cluster 6, it has a different time variability as compared with the rest of the resolved clusters with contributions peaking in the afternoon and increasing in the fall and winter seasons. Therefore, they do not come from Santiago's lower valley. RM results [61] indicate that regional anthropogenic sources contribute to ambient $PM_{2.5}$ at the monitoring site, diagnosed by the presence of arsenic and sulfur in ambient $PM_{2.5}$ samples. To check whether cluster 6 could represent those regional sources, Figure S6 shows the source contributions to ambient SO_2. It can be seen that cluster 6 SO_2 contributions show up all year long, so they cannot come from Santiago's lower valley. Besides, Figure S5 shows that cluster 6 has W and WSW wind directions, suggesting that those regional sources are located west of Santiago. This result agrees with the location of regional sources of arsenic and sulfates identified in Reference [61] using backward trajectory analyses. Therefore, we conclude that cluster 6 can be identified as representative of regional sources of $PM_{2.5}$.

In order to make comparisons between the CA results and the RM results, we need to consider the limitations of both analyses. The RM results were obtained using Positive Matrix Factorization software and elements as tracers. Organic tracers were not included and this is a limitation. For instance, more recent results [90,91] have shown that secondary organic aerosols are relevant in the spring and summer seasons in Santiago. This secondary $PM_{2.5}$ was not resolved by the PMF solution including only elements. Another limitation of the PMF solution is the use of potassium as a tracer of wood burning. Potassium may be a reasonable tracer in fall and winter seasons, but not so specific in the spring and summer when soil dust becomes relevant in Santiago's semi-arid climate [70]. Because of these issues, we have decided to compare CA and RM results only for the months from May through August 2004. Table 4 shows such a comparison. We have grouped clusters 3 and 4 as traffic sources, cluster 7 as the RWB source, and cluster 6 as regional sources. The smaller contributions from other clusters have not been considered, given the limitations of both CA and RM to resolve smaller (or intermittent) sources. Since the filter-based RM results consider only a subset of days per month, these specific days have been extracted from the CA results to compute comparable averages.

Table 4. Ambient $PM_{2.5}$ source apportionment (±standard error), Santiago, May–August 2004 ($\mu g/m^3$).

Month	Traffic (CA)	Traffic (RM)	RWB (CA)	RWB (RM)	Regional (CA)	Regional (RM)
May	25.2 ± 2.5	14.7 ± 1.2	7.4 ± 2.4	15.1 ± 1.6	6.3 ± 1.2	17.7 ± 2.2
June	20.8 ± 1.9	17.6 ± 1.7	6.4 ± 2.1	13.8 ± 1.2	7.9 ± 1.3	11.6 ± 1.1
July	21.6 ± 2.1	16.8 ± 1.5	4.0 ± 1.6	12.7 ± 1.3	6.7 ± 1.7	12.7 ± 1.7
August	14.8 ± 1.8	9.3 ± 0.8	8.1 ± 2.3	7.3 ± 0.7	4.0 ± 1.1	12.3 ± 1.3

From the results in Table 4, it can be seen that CA tends to overestimate the RM result for traffic contributions (which includes traffic and soil dust contributions) but CA results for RWB and regional sources contributions that are below those estimated by the RM in the very same monitoring site. In fact, the sum of CA estimates in Table 3 has an average of 33.3 ($\mu g/m^3$), while the corresponding average of RM results is 41.3 ($\mu g/m^3$). The reason for this discrepancy is ascribed to the different $PM_{2.5}$ measurement techniques: filter samples were taken in low-volume dichotomous samplers (Andersen Instrument, Inc, Smyrna, GA, USA, 15 L/min) while continuous measurements were made with a Tapered Element Oscillating Microbalance equipment (TEOM, Rupprecht & Patashnick, MA, USA). The latter instrument is known to present negative artifacts (i.e., underestimation of $PM_{2.5}$ concentrations) due to partial volatilization of sampled $PM_{2.5}$ in the TEOM's heating inlet used to dry the samples [92]. We think this instrument artifact explains why filter-based RM results are higher than the CA results reported here. This effect explains the lower contributions found for RWB in the CA because RWB sources have a larger fraction of volatile compounds emitted, as compared with other combustion sources [53]. We do not have such an issue in the case of Temuco because, in that monitoring site, a beta-attenuation monitor (BAM, MET-ONE 1020, Met One Instruments Inc., Grants Pass, OR, USA) has been used to measure $PM_{2.5}$ and PM_{10}.

4. Conclusions

We have proposed that a CA approach, applied to ambient air pollution and meteorological data, provides a source apportionment of ambient $PM_{2.5}$ and PM_{10} on an hourly basis. In order to achieve this result, we need to compare the outcome of the CA with RM results for the same monitoring site—or using an emission inventory in case no RM result is available—to identify the major sources (clusters) at play on a given zone. We have shown that our rule-based CA works for three different case studies in Chile: a city in a warm, desert region (Calama), another one dominated by RWB pollution in a cold, wet region (Temuco), and a large metropolitan area in a semi-arid region (Santiago).

In the case of Calama, the CA for ambient PM_{10} is able to resolve local sources (traffic) and windblown dust coming from the nearby desert environment. Both are fugitive sources, which are difficult to estimate, because of the amount of information required on each case, such as particle size distributions. CA results indicate that traffic sources dominate ambient PM_{10} concentrations, as suggested by the emission inventory of PM_{10} sources for that city. CA results show that, over the long-term, outbursts of windblown dust are not significant within the city. This is relevant for policy purposes. The evolution of traffic contributions in Calama (Figure 3) suggests that the ongoing street sweeping program has been a successful one with a clear decreasing trend. This is an example of how a sector-specific regulation in the city may be evaluated. Furthermore, within the traffic source, ambient PM_{10} data can be regarded as a combination of exhaust and non-exhaust emissions, providing additional insights such as whether new exhaust emission standards for motor vehicles have curbed down exhaust emissions.

In the case of Temuco, with dominant RWB contributions to ambient $PM_{2.5}$, the use of ambient temperature instead of wind speed improves the apportionment of all major sources of ambient $PM_{2.5}$. RWB contributions show up in the bivariate polar plots as a 'central' cluster because, under stable atmospheric conditions (with lowest ambient temperatures and wind speeds), wind direction is highly variable. Hence, the monitor site will sample air masses from all wind directions. CA source contributions for RWB sources are statistically comparable with RM results obtained for RWB sources in a short-term campaign in 2014.

For the large metropolitan area of Santiago, the CA methodology resolved RWB, traffic, and regional sources of ambient $PM_{2.5}$. The CA analysis again required using ambient temperature to resolve the RWB contribution. The seasonality of the resolved clusters showed a distinctive effect brought by topography: Santiago's plume does not reach the eastern side of the city in fall and winter seasons due to low PBL depths in colder months. This geographic condition provided an additional criterium to identify local and non-local traffic sources. However, the RM resolved source contributions (short-term campaign in 2004) were higher than the CA resolved counterparts. This is a result of an underestimation of ambient $PM_{2.5}$ monitoring brought by volatilization losses in the continuous TEOM $PM_{2.5}$ monitor.

The rule-based CA presented here generates added value to existing ambient air pollution databases. Regarding RM results to complement the CA analysis, methods that use specific source tracers (like organic molecular markers) are preferable.

The rule-based CA results may be applied to the following analyses:

1. Identifying specific meteorological conditions leading to high $PM_{2.5}$ concentrations, like windblown dust in arid regions.
2. Provide long-term time series of source contributions to constrain emissions through DM applications to improve emission inventories.
3. Tracking source trends and assess efficiency of specific regulations.
4. Conduct environmental justice studies with the aid of low-cost air pollution monitoring (citizen science).
5. Identify intermittent sources contributions, which may be further pinpointed using backward trajectory analysis.
6. Conduct epidemiological studies to find associations between health effects and exposure to a single $PM_{2.5}$ source such as traffic.
7. Help in analyzing massive databases coming from state-of-the-science continuous monitors such as time-of-flight mass spectrometers measuring aerosols or VOC, multi-wavelength aethalometers, etc.

The proposed rule-based CA analysis has limitations, though. The methodology works well for resolving the larger sources at play in a city, but smaller sources remain a challenge.

Supplementary Materials: The following are available online at http://www.mdpi.com/1660-4601/17/22/8455/s1. Table S1: Emission inventory for Calama, year 2016 [ton/year]. Figure S1: PM_{10} Time variability results for a five-cluster solution for Calama. Figure S2: PM_{10}-wind speed scatter plot by cluster for a 5-cluster solution for Calama. Figure S3: Pollution rose results by cluster for a 3-cluster solution for Temuco. Figure S4: Time variability of CO by cluster for a 3-cluster solution for Temuco. Figure S5: Pollution rose by cluster for an 8-cluster solution for Santiago. Figure S6: Source apportionment for SO_2 for an 8-cluster solution for Santiago.

Author Contributions: Conceptualization, H.J. Methodology, H.J. Validation, A.M.V. Writing—original draft preparation, H.J. Writing—review and editing, H.J. and A.M.V. All authors have read and agreed to the published version of the manuscript.

Funding: This research was funded by Chile's ANID, grant number FONDAP/15110020 (CEDEUS).

Acknowledgments: We acknowledge Gabriela Mancheno for her help in drawing some figures. We also thank two anonymous reviewers for their helpful comments and suggestions.

Conflicts of Interest: The authors declare no conflict of interest. The funders had no role in the design of the study, in the collection, analyses, or interpretation of data, in the writing of the manuscript, or in the decision to publish the results.

References

1. Lelieveld, J.; Evans, J.S.; Fnais, M.; Giannadaki, D.; Pozzer, A. The contribution of outdoor air pollution sources to premature mortality on a global scale. *Nature* **2015**, *525*, 367. [CrossRef]

2. Burnett, R.; Chen, H.; Szyszkowicz, M.; Fann, N.; Hubbell, B.; Pope, C.A.; Apte, J.S.; Brauer, M.; Cohen, A.; Weichenthal, S.; et al. Global estimates of mortality associated with long-term exposure to outdoor fine particulate matter. *Proc. Natl. Acad. Sci. USA* **2018**, *115*, 201803222. [CrossRef]

3. Seinfeld, J.; Pandis, S. *Atmospheric Chemistry and Physics. From Air Pollution to Climate Change*, 2nd ed.; John Wiley & Sons: Hoboken, NJ, USA, 2006; ISBN 978-0471720188.

4. Jimenez, J.L.; Canagaratna, M.R.; Donahue, N.M.; Prevot, A.S.H.; Zhang, Q.; Kroll, J.H.; Decarlo, P.F.; Allan, J.D.; Coe, H.; Ng, N.L.; et al. Evolution of organic aerosols in the atmosphere. *Science* **2009**, *326*, 1525–1529. [CrossRef] [PubMed]

5. Wang, C.; Yuan, T.; Wood, S.; Goss, K.U.; Li, J.; Ying, Q.; Wania, F. Uncertain Henry's law constants compromise equilibrium partitioning calculations of atmospheric oxidation products. *Atmos. Chem. Phys.* **2017**, *17*, 7529–7540. [CrossRef]

6. WHO. WHO Global Urban Ambient Air Pollution Database (update 2016). Available online: http://www. who.int/phe/health_topics/outdoorair/databases/cities/en/ (accessed on 6 April 2018).

7. Cassiani, M.; Stohl, A.; Eckhardt, S. The dispersion characteristics of air pollution from the world's megacities. *Atmos. Chem. Phys.* **2013**, *13*, 9975–9996. [CrossRef]

8. Stock, Z.S.; Russo, M.R.; Butler, T.M.; Archibald, A.T.; Lawrence, M.G.; Telford, P.J.; Abraham, N.L.; Pyle, J.A. Modelling the impact of megacities on local, regional and global tropospheric ozone and the deposition of nitrogen species. *Atmos. Chem. Phys.* **2013**, *13*, 12215–12231. [CrossRef]

9. Liu, Y.; Liu, Y.; Muñoz-Esparza, D.; Hu, F.; Yan, C.; Miao, S. Simulation of Flow Fields in Complex Terrain with WRF-LES: Sensitivity Assessment of Different PBL Treatments. *J. Appl. Meteorol. Climatol.* **2020**, *59*, 1481–1501. [CrossRef]

10. Edwards, J.M.; Beljaars, A.C.M.; Holtslag, A.A.M.; Lock, A.P. Representation of Boundary-Layer Processes in Numerical Weather Prediction and Climate Models. *Boundary-Layer Meteorol.* **2020**. [CrossRef]

11. Simon, H.; Baker, K.R.; Phillips, S. Compilation and interpretation of photochemical model performance statistics published between 2006 and 2012. *Atmos. Environ.* **2012**, *61*, 124–139. [CrossRef]

12. Belis, C.A.; Karagulian, F.; Amato, F.; Almeida, M.; Artaxo, P.; Beddows, D.C.S.; Bernardoni, V.; Bove, M.C.; Carbone, S.; Cesari, D.; et al. A new methodology to assess the performance and uncertainty of source apportionment models II: The results of two European intercomparison exercises. *Atmos. Environ.* **2015**, *123*. [CrossRef]

13. Hopke, P.K. A Review of Receptor Modeling Methods for Source Apportionment. *J. Air Waste Manage. Assoc.* **2016**, *66*, 237–259. [CrossRef] [PubMed]

14. Belis, C.A.; Karagulian, F.; Larsen, B.R.; Hopke, P.K. Critical review and meta-analysis of ambient particulate matter source apportionment using receptor models in Europe. *Atmos. Environ.* **2013**, *69*, 94–108. [CrossRef]

15. Belis, C.A.; Pernigotti, D.; Pirovano, G.; Favez, O.; Jaffrezo, J.L.; Kuenen, J.; Denier van Der Gon, H.; Reizer, M.; Riffault, V.; Alleman, L.Y.; et al. Evaluation of receptor and chemical transport models for PM10 source apportionment. *Atmos. Environ. X* **2020**, *5*, 100053. [CrossRef]

16. Hopke, P.K.; Dai, Q.; Li, L.; Feng, Y. Global review of recent source apportionments for airborne particulate matter. *Sci. Total Environ.* **2020**, *740*, 140091. [CrossRef] [PubMed]

17. Le Breton, M.; Psichoudaki, M.; Hallquist, M.; Watne, K.; Lutz, A.; Hallquist, M. Application of a FIGAERO ToF CIMS for on-line characterization of real-world fresh and aged particle emissions from buses. *Aerosol Sci. Technol.* **2019**, *53*, 244–259. [CrossRef]

18. Holzinger, R.; Goldstein, A.H.; Hayes, P.L.; Jimenez, J.L.; Timkovsky, J. Chemical evolution of organic aerosol in Los Angeles during the CalNex 2010 study. *Atmos. Chem. Phys.* **2013**, *13*, 10125–10141. [CrossRef]

19. Fischer, D.A.; Smith, G.D. A portable, four-wavelength, single-cell photoacoustic spectrometer for ambient aerosol absorption. *Aerosol Sci. Technol.* **2018**, *52*, 393–406. [CrossRef]

20. Massabò, D.; Bernardoni, V.; Bove, M.C.; Brunengo, A.; Cuccia, E.; Piazzalunga, A.; Prati, P.; Valli, G.; Vecchi, R. A multi-wavelength optical set-up for the characterization of carbonaceous particulate matter. *J. Aerosol Sci.* **2013**, *60*, 34–46. [CrossRef]

21. Massabò, D.; Caponi, L.; Bernardoni, V.; Bove, M.C.; Brotto, P.; Calzolai, G.; Cassola, F.; Chiari, M.; Fedi, M.E.; Fermo, P.; et al. Multi-wavelength optical determination of black and brown carbon in atmospheric aerosols. *Atmos. Environ.* **2015**, *108*, 1–12. [CrossRef]

22. Miller, A.J.; Raduma, D.M.; George, L.A.; Fry, J.L. Source apportionment of trace elements and black carbon in an urban industrial area (Portland, Oregon). *Atmos. Pollut. Res.* **2019**, *10*, 784–794. [CrossRef]

23. Küpper, M.; Quass, U.; John, A.C.; Kaminski, H.; Leinert, S.; Breuer, L.; Gladtke, D.; Weber, S.; Kuhlbusch, T.A.J. Contributions of carbonaceous particles from fossil emissions and biomass burning to PM10 in the Ruhr area, Germany. *Atmos. Environ.* **2018**, *189*, 174–186. [CrossRef]

24. Helin, A.; Niemi, J.V.; Virkkula, A.; Pirjola, L.; Teinilä, K.; Backman, J.; Aurela, M.; Saarikoski, S.; Rönkkö, T.; Asmi, E.; et al. Characteristics and source apportionment of black carbon in the Helsinki metropolitan area, Finland. *Atmos. Environ.* **2018**, *190*, 87–98. [CrossRef]

25. Boleti, E.; Hueglin, C.; Takahama, S. Trends of surface maximum ozone concentrations in Switzerland based on meteorological adjustment for the period 1990–2014. *Atmos. Environ.* **2019**, *213*, 326–336. [CrossRef]

26. Gurjar, B.R.; Ravindra, K.; Nagpure, A.S. Air pollution trends over Indian megacities and their local-to-global implications. *Atmos. Environ.* **2016**, *142*, 475–495. [CrossRef]

27. Faridi, S.; Shamsipour, M.; Krzyzanowski, M.; Künzli, N.; Amini, H.; Azimi, F.; Malkawi, M.; Momeniha, F.; Gholampour, A.; Hassanvand, M.S.; et al. Long-term trends and health impact of PM2.5 and O3 in Tehran, Iran, 2006–2015. *Environ. Int.* **2018**, *114*, 37–49. [CrossRef]

28. Wu, Z.; Zhang, Y.; Zhang, L.; Huang, M.; Zhong, L.; Chen, D.; Wang, X. Trends of outdoor air pollution and the impact on premature mortality in the Pearl River Delta region of southern China during 2006–2015. *Sci. Total Environ.* **2019**, *690*, 248–260. [CrossRef]

29. Carvalho, V.S.B.; Freitas, E.D.; Martins, L.D.; Martins, J.A.; Mazzoli, C.R.; Andrade, M. de F. Air quality status and trends over the Metropolitan Area of São Paulo, Brazil as a result of emission control policies. *Environ. Sci. Policy* **2015**, *47*, 68–79. [CrossRef]

30. McKercher, G.R.; Salmond, J.A.; Vanos, J.K. Characteristics and applications of small, portable gaseous air pollution monitors. *Environ. Pollut.* **2017**, *223*, 102–110. [CrossRef]

31. Jovašević-Stojanović, M.; Bartonova, A.; Topalović, D.; Lazović, I.; Pokrić, B.; Ristovski, Z. On the use of small and cheaper sensors and devices for indicative citizen-based monitoring of respirable particulate matter. *Environ. Pollut.* **2015**, *206*, 696–704. [CrossRef]

32. Feinberg, S.N.; Williams, R.; Hagler, G.; Low, J.; Smith, L.; Brown, R.; Garver, D.; Davis, M.; Morton, M.; Schaefer, J.; et al. Examining spatiotemporal variability of urban particulate matter and application of high-time resolution data from a network of low-cost air pollution sensors. *Atmos. Environ.* **2019**, *213*, 579–584. [CrossRef]

33. Sayahi, T.; Butterfield, A.; Kelly, K.E. Long-term field evaluation of the Plantower PMS low-cost particulate matter sensors. *Environ. Pollut.* **2019**, *245*, 932–940. [CrossRef] [PubMed]

34. Weissert, L.; Alberti, K.; Miles, E.; Miskell, G.; Feenstra, B.; Henshaw, G.S.; Papapostolou, V.; Patel, H.; Polidori, A.; Salmond, J.A.; et al. Low-cost sensor networks and land-use regression: Interpolating nitrogen dioxide concentration at high temporal and spatial resolution in Southern California. *Atmos. Environ.* **2020**, *223*, 117287. [CrossRef]

35. van Zoest, V.; Osei, F.B.; Stein, A.; Hoek, G. Calibration of low-cost NO2 sensors in an urban air quality network. *Atmos. Environ.* **2019**, *210*, 66–75. [CrossRef]

36. Johnson, K.K.; Bergin, M.H.; Russell, A.G.; Hagler, G.S.W. Using Low Cost Sensors to Measure Ambient Particulate Matter Concentrations and On-Road Emissions Factors. *Atmos. Meas. Tech. Discuss.* **2016**, 1–22. [CrossRef]

37. Miskell, G.; Alberti, K.; Feenstra, B.; Henshaw, G.S.; Papapostolou, V.; Patel, H.; Polidori, A.; Salmond, J.A.; Weissert, L.; Williams, D.E. Reliable data from low cost ozone sensors in a hierarchical network. *Atmos. Environ.* **2019**, *214*, 116870. [CrossRef]

38. Grange, S.K.; Lewis, A.C.; Carslaw, D.C. Source apportionment advances using polar plots of bivariate correlation and regression statistics. *Atmos. Environ.* **2016**, *145*, 128–134. [CrossRef]

39. Uria-Tellaetxe, I.; Carslaw, D.C. Conditional bivariate probability function for source identification. *Environ. Model. Softw.* **2014**, *59*, 1–9. [CrossRef]
40. Carslaw, D.C.; Beevers, S.D. Characterising and understanding emission sources using bivariate polar plots and k-means clustering. *Environ. Model. Softw.* **2013**, *40*, 325–329. [CrossRef]
41. Govender, P.; Sivakumar, V. Application of k-means and hierarchical clustering techniques for analysis of air pollution: A review (1980–2019). *Atmos. Pollut. Res.* **2020**, *11*, 40–56. [CrossRef]
42. Austin, E.; Coull, B.; Thomas, D.; Koutrakis, P. A framework for identifying distinct multipollutant profiles in air pollution data. *Environ. Int.* **2012**, *45*, 112–121. [CrossRef]
43. Karagulian, F.; Belis, C.A.; Dora, C.F.C.; Pruss-Ustun, A.M.; Bonjour, S.; Adair-Rohani, H.; Amann, M. Contributions to cities' ambient particulate matter (PM): A systematic review of local source contributions at global level. *Atmos. Environ.* **2015**, *120*, 475–483. [CrossRef]
44. Mukherjee, A.; Agrawal, M. World air particulate matter: Sources, distribution and health effects. *Environ. Chem. Lett.* **2017**, *15*, 283–309. [CrossRef]
45. Amato, F.; Cassee, F.R.; Denier van der Gon, H.A.C.; Gehrig, R.; Gustafsson, M.; Hafner, W.; Harrison, R.M.; Jozwicka, M.; Kelly, F.J.; Moreno, T.; et al. Urban air quality: The challenge of traffic non-exhaust emissions. *J. Hazard. Mater.* **2014**, *275*, 31–36. [CrossRef] [PubMed]
46. Amato, F.; Nava, S.; Lucarelli, F.; Querol, X.; Alastuey, A.; Baldasano, J.M.; Pandolfi, M. A comprehensive assessment of PM emissions from paved roads: Real-world Emission Factors and intense street cleaning trials. *Sci. Total Environ.* **2010**, *408*, 4309–4318. [CrossRef]
47. Kalaiarasan, G.; Balakrishnan, R.M.; Sethunath, N.A.; Manoharan, S. Source apportionment studies on particulate matter (PM10 and PM2.5) in ambient air of urban Mangalore, India. *J. Environ. Manag.* **2018**, *217*, 815–824. [CrossRef] [PubMed]
48. Crilley, L.R.; Lucarelli, F.; Bloss, W.J.; Harrison, R.M.; Beddows, D.C.; Calzolai, G.; Nava, S.; Valli, G.; Bernardoni, V.; Vecchi, R. Source apportionment of fine and coarse particles at a roadside and urban background site in London during the 2012 summer ClearfLo campaign. *Environ. Pollut.* **2017**, *220*, 766–778. [CrossRef]
49. Pant, P.; Harrison, R.M. Estimation of the contribution of road traffic emissions to particulate matter concentrations from field measurements: A review. *Atmos. Environ.* **2013**, *77*, 78–97. [CrossRef]
50. Grigoratos, T.; Martini, G. Brake wear particle emissions: A review. *Environ. Sci. Pollut. Res.* **2015**, *22*, 2491–2504. [CrossRef]
51. Thorpe, A.; Harrison, R.M. Sources and properties of non-exhaust particulate matter from road traffic: A review. *Sci. Total Environ.* **2008**, *400*, 270–282. [CrossRef]
52. May, A.A.; Presto, A.A.; Hennigan, C.J.; Nguyen, N.T.; Gordon, T.D.; Robinson, A.L. Gas-particle partitioning of primary organic aerosol emissions: (1) Gasoline vehicle exhaust. *Atmos. Environ.* **2013**, *77*, 128–139. [CrossRef]
53. Robinson, A.L.; Grieshop, A.P.; Donahue, N.M.; Hunt, S.W. Updating the Conceptual Model for Fine Particle Mass Emissions from Combustion Systems Allen L. Robinson. *J. Air Waste Manage. Assoc.* **2010**, *60*, 1204–1222. [CrossRef] [PubMed]
54. Roden, C.A.; Bond, T.C.; Conway, S.; Osorto Pinel, A.B. Emission factors and real-time optical properties of particles emitted from traditional wood burning cookstoves. *Environ. Sci. Technol.* **2006**, *40*, 6750–6757. [CrossRef] [PubMed]
55. Bray, C.D.; Strum, M.; Simon, H.; Riddick, L.; Kosusko, M.; Menetrez, M.; Hays, M.D.; Rao, V. An assessment of important SPECIATE profiles in the EPA emissions modeling platform and current data gaps. *Atmos. Environ.* **2019**, *207*, 93–104. [CrossRef] [PubMed]
56. Corbin, J.C.; Keller, A.; Lohmann, U.; Burtscher, H.; Sierau, B.; Mensah, A.A. Organic Emissions from a Wood Stove and a Pellet Stove Before and After Simulated Atmospheric Aging. *Aerosol Sci. Technol.* **2015**, *49*, 1037–1050. [CrossRef]
57. Bruns, E.A.; Krapf, M.; Orasche, J.; Huang, Y.; Zimmermann, R.; Drinovec, L.; Močnik, G.; El-Haddad, I.; Slowik, J.G.; Dommen, J.; et al. Characterization of primary and secondary wood combustion products generated under different burner loads. *Atmos. Chem. Phys.* **2015**, *15*, 2825–2841. [CrossRef]
58. Lopez-Aparicio, S.; Grythe, H. Evaluating the effectiveness of a stove exchange programme on PM2.5 emission reduction. *Atmos. Environ.* **2020**, *231*, 117529. [CrossRef]

59. Carslaw, D.C.; Ropkins, K. Openair - An r package for air quality data analysis. *Environ. Model. Softw.* **2012**, *27–28*, 52–61. [CrossRef]

60. Villalobos, A.M.; Barraza, F.; Jorquera, H.; Schauer, J.J. Wood burning pollution in southern Chile: PM2.5 source apportionment using CMB and molecular markers. *Environ. Pollut.* **2017**, *225*, 514–523. [CrossRef]

61. Jorquera, H.; Barraza, F. Source apportionment of ambient PM 2.5 in Santiago, Chile: 1999 and 2004 results. *Sci. Total Environ.* **2012**, *435–436*, 418–429. [CrossRef]

62. U.S. EPA Positive Matrix Factorization, Version 5. Available online: https://www.epa.gov/air-research/positive-matrix-factorization-model-environmental-data-analyses (accessed on 31 October 2020).

63. U.S. EPA Chemical Mass Balance Software (CMB) Version 8.2. Available online: https://www.epa.gov/scram/chemical-mass-balance-cmb-model (accessed on 31 October 2020).

64. Watson, J.G.; Cooper, J.A.; Huntzicker, J.J. The effective variance weighting for least squares calculations applied to the mass balance receptor model. *Atmos. Environ.* **1984**, *18*, 1347–1355. [CrossRef]

65. Henry, R.C. History and fundamentals of multivariate air quality receptor models. *Chemom. Intell. Lab. Syst.* **1997**, *37*, 37–42. [CrossRef]

66. INE Chile's Population Census 2017. Available online: https://www.ine.cl/estadisticas/sociales/demografia-y-vitales/proyecciones-de-poblacion (accessed on 31 October 2020).

67. Kottek, M.; Grieser, J.; Beck, C.; Rudolf, B.; Rubel, F. World Map of the Köppen-Geiger climate classification updated. *Meteorol. Zeitschrift* **2006**, *15*, 259–263. [CrossRef]

68. Jhun, I.; Oyola, P.; Moreno, F.; Castillo, M.A.; Koutrakis, P. PM2.5 mass and species trends in Santiago, Chile, 1998 to 2010: The impact of fuel-related interventions and fuel sales. *J. Air Waste Manage. Assoc.* **2013**, *63*, 161–169. [CrossRef] [PubMed]

69. Jorquera, H.; Orregom, G.; Castro, J.; Vesovic, V.; Orrego, G.; Castro, J.; Vesovic, V. Trends in air quality and population exposure in Santiago, Chile, 1989-2001. *Int. J. Environ. Pollut.* **2004**, *22*, 507–530. [CrossRef]

70. Barraza, F.; Lambert, F.; Jorquera, H.; Villalobos, A.M.; Gallardo, L. Temporal evolution of main ambient PM2.5 sources in Santiago, Chile, from 1998 to 2012. *Atmos. Chem. Phys.* **2017**, *17*, 10093–10107. [CrossRef]

71. Jorquera, H. Ambient particulate matter in Santiago, Chile: 1989–2018: A tale of two size fractions. *J. Environ. Manage.* **2020**, *258*, 110035. [CrossRef]

72. Jorquera, H.; Barraza, F.; Heyer, J.; Valdivia, G.; Schiappacasse, L.N.; Montoya, L.D. Indoor PM2.5in an urban zone with heavy wood smoke pollution: The case of Temuco, Chile. *Environ. Pollut.* **2018**, *236*, 477–487. [CrossRef]

73. Jorquera, H.; Barraza, F. Source apportionment of PM 10 and PM 2. 5 in a desert region in northern Chile. *Sci. Total Environ.* **2013**, *444*, 327–335. [CrossRef]

74. Grange, S.K.; Carslaw, D.C. Using meteorological normalisation to detect interventions in air quality time series. *Sci. Total Environ.* **2019**, *653*, 578–588. [CrossRef]

75. Henry, R.C. Multivariate receptor modeling by N-dimensional edge detection. *Chemom. Intell. Lab. Syst.* **2003**, *65*, 179–189. [CrossRef]

76. Hassan, H.; Saraga, D.; Kumar, P.; Kakosimos, K.E. Vehicle-induced fugitive particulate matter emissions in a city of arid desert climate. *Atmos. Environ.* **2020**, *229*, 117450. [CrossRef]

77. Lupu, A.; Maenhaut, W. Application and comparison of two statistical trajectory techniques for identification of source regions of atmospheric aerosol species. *Atmos. Environ.* **2002**, *36*, 5607–5618. [CrossRef]

78. Pérez, I.A.; Artuso, F.; Mahmud, M.; Kulshrestha, U.; Sánchez, M.L.; García, M.Á. Applications of air mass trajectories. *Adv. Meteorol.* **2015**, *2015*. [CrossRef]

79. WHO Air Quality Guidelines—Global Update 2005. Available online: http://www.who.int/phe/health_topics/outdoorair/outdoorair_aqg/en/ (accessed on 29 September 2020).

80. DICTUC Antecedentes Tecnicos Para el PDA de Calama. Available online: http://planesynormas.mma.gob.cl/archivos/2019/proyectos/25052019_dictuc26ab19.rar (accessed on 29 September 2020).

81. Rutllant, J.A. Climate dynamics along the arid northern coast of Chile: The 1997–1998 Dinámica del Clima de la Región de Antofagasta (DICLIMA) experiment. *J. Geophys. Res.* **2003**, *108*, 4538. [CrossRef]

82. Turpin, B.J.; Lim, H.-J. Species Contributions to PM2.5 Mass Concentrations: Revisiting Common Assumptions for Estimating Organic Mass. *Aerosol Sci. Technol.* **2001**, *35*, 602–610. [CrossRef]

83. Sheesley, R.J.; Schauer, J.J.; Zheng, M.; Wang, B. Sensitivity of molecular marker-based CMB models to biomass burning source profiles. *Atmos. Environ.* **2007**, *41*, 9050–9063. [CrossRef]

84. Rogge, W.F.; Hildemann, L.M.; Mazurek, M.A.; Cass, G.R.; Simoneit, B.R.T. Sources of Fine Organic Aerosol. 4. Particulate Abrasion Products From Leaf Surfaces of Urban Plants. *Environ. Sci. Technol.* **1993**, *27*, 2700–2711. [CrossRef]
85. Zhang, Y.; Schauer, J.J.; Zhang, Y.; Zeng, L.; Wei, Y.; Liu, Y.; Shao, M. Characteristics of particulate carbon emissions from real-world Chinese coal combustion. *Environ. Sci. Technol.* **2008**, *42*, 5068–5073. [CrossRef]
86. Nalin, F.; Golly, B.; Besombes, J.-L.; Pelletier, C.; Aujay-Plouzeau, R.; Verlhac, S.; Dermigny, A.; Fievet, A.; Karoski, N.; Dubois, P.; et al. Fast oxidation processes from emission to ambient air introduction of aerosol emitted by residential log wood stoves. *Atmos. Environ.* **2016**, *143*, 15–26. [CrossRef]
87. Gramsch, E.; Muñoz, A.; Langner, J.; Morales, L.; Soto, C.; Pérez, P.; Rubio, M.A. Black carbon transport between Santiago de Chile and glaciers in the Andes Mountains. *Atmos. Environ.* **2020**, *232*, 117546. [CrossRef]
88. Muñoz, R.C.; Undurraga, A.A. Daytime mixed layer over the Santiago Basin: Description of two years of observations with a lidar ceilometer. *J. Appl. Meteorol. Climatol.* **2010**, *49*, 1728–1741. [CrossRef]
89. Gramsch, E.; Cereceda-Balic, F.; Oyola, P.; von Baer, D. Examination of pollution trends in Santiago de Chile with cluster analysis of PM10 and Ozone data. *Atmos. Environ.* **2006**, *40*, 5464–5475. [CrossRef]
90. Carbone, S.; Saarikoski, S.; Frey, A.; Reyes, F.; Reyes, P.; Castillo, M.; Gramsch, E.; Oyola, P.; Jayne, J.; Worsnop, D.; et al. Chemical characterization of submicron Aerosol particles in Santiago de Chile. *Aerosol Air Qual. Res.* **2013**, *13*, 462–473. [CrossRef]
91. Villalobos, A.M.; Barraza, F.; Jorquera, H.; Schauer, J.J. Chemical speciation and source apportionment of fine particulate matter in Santiago, Chile, 2013. *Sci. Total Environ.* **2015**, *512–513*. [CrossRef]
92. Li, Q.F.; Wang-Li, L.; Liu, Z.; Heber, A.J. Field evaluation of particulate matter measurements using tapered element oscillating microbalance in a layer house. *J. Air Waste Manag. Assoc.* **2012**, *62*, 322–335. [CrossRef]

Publisher's Note: MDPI stays neutral with regard to jurisdictional claims in published maps and institutional affiliations.

International Journal of
*Environmental Research
and Public Health*

Article

Analysis of Alpha Activity Levels and Dependence on Meteorological Factors over a Complex Terrain in Northern Iberian Peninsula (2014–2018)

Miguel Ángel Hernández-Ceballos [1], **Fernando Legarda** [2] **and Natalia Alegría** [2,*]

1   Department of Physics, University of Cordoba, 14071 Córdoba, Spain; f92hecem@uco.es
2   Department of Energetic Engineering, University of the Basque Country, 48013 Bilbao, Spain; f.legarda@ehu.eus
*   Correspondence: Natalia.alegria@ehu.eus; Tel.: +34-94-601-7279

Received: 18 September 2020; Accepted: 28 October 2020; Published: 29 October 2020

Abstract: Alpha ambient concentrations in ground-level air were measured weekly in Bilbao (northern Spain) by collecting aerosols in filters between 2014 and 2018. Over this period, the alpha activity concentrations in the aerosol's samples range from 13.9 $\mu Bq/m^3$ to 246.5 $\mu Bq/m^3$, with a mean of 66.49 ± 39.33 $\mu Bq/m^3$. The inter-annual and intra-annual (seasonal and monthly) variations are analyzed, with the highest activity in autumn months and the lowest one in winter months. Special attention has been paid to alpha peak concentrations (weekly concentrations above the 90th percentile) and its relationship with regional meteorological scenarios by means of air mass trajectories and local meteorological parameters. The meteorological analysis of these high alpha concentrations has revealed two airflow patterns-one from the south with land origin and one from the north with maritime origin-mainly associated with these alpha peak concentrations. Surface winds during representative periods of both airflow patterns are also analyzed in combination with ^{222}Rn concentrations, which demonstrated the different daily evolution associated with each airflow pattern. The present results are relevant in understanding trends and meteorological factors affecting alpha activity concentrations in this area, and hence, to control potential atmospheric environmental releases and ensure the environmental and public health.

Keywords: gross alpha activity; northern Iberian Peninsula; radon; airflow patterns; surface winds

1. Introduction

Radioactivity is naturally present in the air due to cosmogenic radionuclides (14C, 7Be, and 3H), cosmic radiation, and natural radionuclides present in all rocks and soils (^{238}U and progenies, ^{232}Th and progenies, and ^{40}K), and its natural background is increased by anthropogenic sources, such as nuclear/radiological accident, nuclear weapon tests, or the production and application of phosphorus fertilizers [1].

The impact of radioactivity on the environment and population health makes the control of radioactivity levels necessary. In this sense, alpha activity concentration is usually used as a control indicator of the radioactivity levels in air, and it is a common measurement of detecting activity peaks associated with radioactive released to the atmosphere [2]. This activity concentration can be measured counting the number of alpha emissions produced per unit of time and air volume (gross alpha), which are determined from the sampling of atmospheric aerosols. The determination and formation of radioactive aerosols in the atmosphere air are analyzed in detail in Papastefanou [3]. In case of alpha activity concentration in aerosols, their origin can be natural due to radon daughters [3].

The long-term monitoring of gross alpha activity concentrations enables us to study trends and periodic variations and, in combination with meteorological parameters, to identify the main meteorological causes influencing the spatial and temporal variability of the environmental radioactivity levels [4]. The seasonal variation and its link with meteorological parameters such as rainfall, temperature, relative humidity, pressure, visibility, wind speed, and wind direction, which influence the concentration of aerosols radiotracers in the atmosphere, has been analyzed in several sites in Spain. Dueñas [5,6] and Piñero-García [7] in the south, García-Talavera [8] in the west, and Saéz-Muñoz [9] in the east provide information about the spatial distribution and the meteorological conditions driven gross alpha concentrations in surface air in Spain. However, in the northern part of Spain, there are completely different meteorological and climatic conditions, and to the authors' knowledge, there are no studies analyzing alpha concentrations.

The direct relevance of alpha activity on the effective dose causes the need of analyzing and characterizing these concentrations. Depending on the aerosol's size, in the respiration process, some aerosol settling occurs in the alveolar spaces [10], and if there are some radionuclides on them, internal irradiation is produced [11]. Due to the inhalation of the short-lived radon decay products, this radionuclide delivers the largest proportion of the natural radiation dose to humans [3]. The alpha energy released by Rn progeny is close to 35 GeV/Bq.

In this paper, we analyze the temporal variation of the radioactive aerosol concentrations by measuring weekly the gross alpha activity over five years (2014–2018) at Bilbao (northern Spain), and due to the potential health impact of alpha peak concentrations, we investigated the meteorological factors controlling them. To this end, the link between local/regional meteorological condition and alpha peak concentrations are examined by means of the calculation of backward trajectories (regional scale) and the temporal variability of surface winds and ^{222}Rn measured (local scale). This last analysis is based on the fact that, in normal situations, gross alpha origin in air is mainly explained due to the presence of long-lived daughters of gaseous ^{222}Rn that are attached to aerosols after cluster formation [3].

The following research issues are addressed in the present study:

1. Analysis of the temporal variation of alpha activity concentrations in surface air;
2. Dependence of alpha peak activity concentrations on airflow patterns;
3. Analysis of the daily temporal evolution of alpha activity concentrations by means the link between surface wind patterns and ^{222}Rn concentrations.

The structure of this paper is the following: Section 2 describes the measurement place, the radioactivity measurements, and the meteorological parameters and data used for the study. Section 3 presents, in a first step, the statistical analysis of the alpha activity concentration, while in a second step, the correlation of high alpha concentrations with regional and local meteorological parameters is investigated. To finalize this results section, the dependence of alpha activity concentrations, by means of ^{222}Rn values, on the surface wind conditions is analyzed. In Section 4, the conclusions are shown.

2. Materials and Methods

2.1. Study Area

Bilbao City (43.26 N, −2.94 W) is placed in the Basque Country region (northern Spain) (Figure 1). Bilbao is about 16 km away from the sea, in a mountainous coastal area at the Gulf of Biscay, and in the narrow valley of the Nervion river. The mountains surrounding Bilbao have low altitude (300–800 m), although 40 km away there are mountains with altitudes of 1500 m.

The climate in Bilbao is warm and temperate, with a significant amount of rainfall during the year. Air masses mainly arrive from the west and north to this area. In the period covered by the present study (2014–2018), the temperature averaged 16.0 °C, while the precipitation was about 980 mm per year.

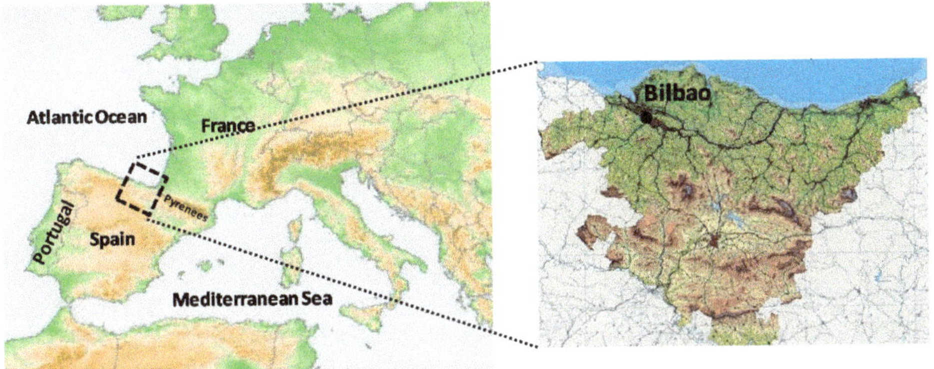

Figure 1. Location of the sampling site in the northern area of the Iberian Peninsula.

2.2. Alpha Activity Concentration and Radon Measurements.

The concentration of radionuclides in soil in this area is very low based on the European Digital Atlas of Natural Radiation [12] (for ^{40}K, uranium, thorium, etc.).

The alpha activity concentrations were measured in a monitoring station sited in Bilbao city (43.26 N, −2.94 W) at 34 m above sea level, on the roof of the Faculty of Engineering (Figure 2).

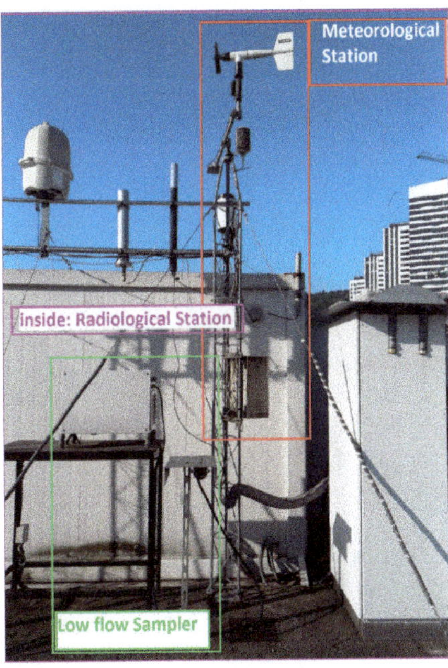

Figure 2. The aerosol sampler, the meteorological station and the radon station used in the present analysis.

Atmospheric airborne aerosols were collected using a low volume sampler whose nominal flow rate is 30 L/min. It is important to remark that the aerosol sampling takes seven days. In a week the total volume circulated is about 300 m^3. This sampler uses a cellulose nitrate membrane filter (diameter 47 mm and 0.8 um pore size) which was replaced weekly. The filter is taken from the sampler and placed in a dryer for six days to allow thoron and radon short-lived alpha progeny to decay, and, then, it is measured in a gas flow proportional counter. The gas flow proportional counter used is LB 770 10-Channel α-β Low-Level Counter by Berthold Technologies GmbH & Co. KG trademark. This device is equipped with 10 detectors. The efficiency of the proportional counter calibrated with [241]Am is close to 20%. A calibration source was prepared ad hoc with the same type of filter as used. The counting time was 6×10^4 seconds, and the detection limits obtained were between 9.71 μBq/m^3 and 50.68 μBq/m^3 and uncertainties, with a coverage factor k = 2, were between 14 $\pm$ 6.7 μBq/m^3.

A high flow sampler and a radiological station are located close to this low flow sampler (Figure 2). The radiological station provides hourly radon activity concentrations in the atmosphere, which have also been used in this study.

2.3. Meteorology and Backward Trajectories

Ten-minute intervals of the following surface meteorological parameters, which were measured in the weather station placed close to the sampling station (Figure 2), were also used in the present analysis to characterize the impact of local meteorology on alpha activity concentrations: air temperature ($^\circ$C), relative humidity (%), pressure (mbar), wind direction ($^\circ$), wind speed (m/s), and precipitation (mm).

Backward trajectories were calculated and used in the present study. Four kinematic three-dimensional backward trajectories per day during specific sampling periods between 2014 and 2018, computed at 00:00, 06:00, 12:00, and 18:00 UTC, with a run time of 96 hours and with a final height of 100 m above ground level, were calculated by using the hybrid single particle Lagrangian integrated trajectory (HYSPLIT) model [13]. The Global Data Assimilation System National Centers for Environmental Prediction (GDAS-NCEP) model, maintained by the National Oceanic and Atmospheric Administration Air Resources Laboratory (NOAA/ARL) were used as meteorological input.

Working with a large number of trajectories requires the application of cluster analysis [14] to classify them according to their pathways, and hence, to identify the corresponding airflow patterns. The present study uses the cluster methodology implemented in the HYSPLIT model and described in Stunder [15]. This multivariate statistical technique groups trajectories with similar direction and lengths by using the Ward's minimum variance clustering method to minimize the dissimilarity in the trajectories within each cluster while maximizing the dissimilarity of different clusters.

3. Results

3.1. Alpha Activity Concentration Characterization: Time Series

The alpha activity concentration values, measured weekly from 2014 to 2018, are plotted in chronological order (Figure 3a). In total, there were 258 data, and they were available in each year, thus guaranteeing the largest statistical sample. The activity concentrations in the aerosol's samples range from 246.5 μBq/m^3 to 13.97 μBq/m^3, with a mean of 66.49 $\pm$ 39.33 μBq/m^3, where the uncertainty is given as the sample standard deviation. This Figure 3a suggests a cyclical variability every year, with low values in the first half of the year followed by an increase and low values at the end of the year. This behavior is only broken in 2015. These alpha activity concentrations are also assessed in a box-and-whisker plot (Figure 3b), which shows that there is a symmetrical distribution of the values, i.e., differences between P75 and P25 with the median (P50) are similar. The average is greater than the P50, which denotes the prevalence of low alpha values, although with large impact of occasional high ones. Using the Grubbs test, the existence of outliers among the registered values has been assessed, concluding that there are no outliers in the present set of data.

Alpha measurements in Figure 3a constitute a time series due to the existence of correlation delays of up to 12 weeks between data [16,17]. The periodogram calculated for this period is shown in Figure 4. A periodogram is a helpful tool used to identify the dominant cyclical behavior of a time series. In the Figure 4, the X-axis represents frequency and the Y-axis represents the spectrum. Four main peaks, i.e., periodicities of alpha concentration records, are identified in Figure 4. The first one reveals a one-year period variation (frequency 1.93×10^{-2} week^{-1} i.e., 52 weeks), while the second, third and fourth peaks indicate intra-annual variability (frequencies 3.9×10^{-2}, 6.9×10^{-2} and 1.6×10^{-1} week^{-1} i.e., 25, 12, and six weeks). These results highly agree with the components of the ^{7}Be time series identified and analyzed in previous studies [4,18]

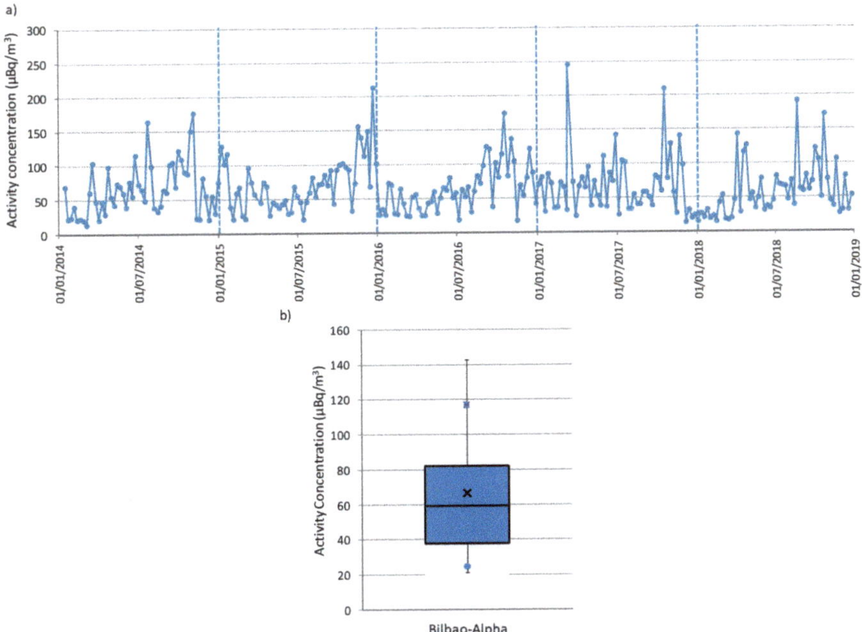

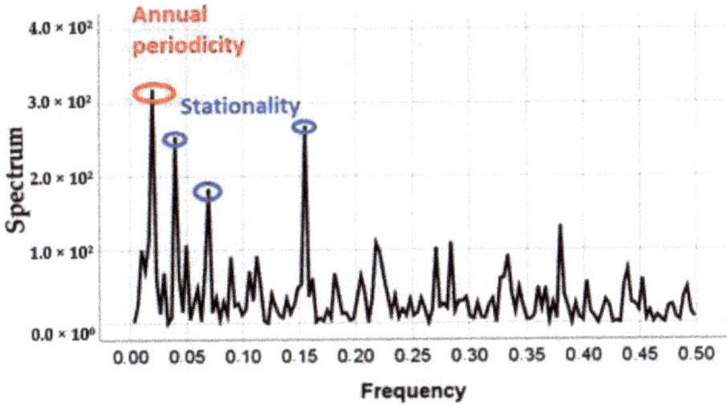

Figure 3. (**a**) Temporal variation and (**b**) box-and-whisker plot of the alpha activity concentrations from January 2014 to December 2018 at Bilbao sampling station. In **b**, the center of the box denotes the 50% (P50), and the bottom and top of the box correspond to the 25% (P25) and 75% (P75) values, respectively. The square indicates the P90 and the circle the P10 values, while the extremes of the box represent the P95 and P5 values. The cross in the box is the average value.

Figure 4. Periodogram for Bilbao time series of alpha activity concentrations from 2014 to 2018.

Figure 5 displays the box-and-whisker plots of alpha activity concentrations over the years. This Figure 5 shows that the average (square) and the variability of alpha values (P75–P25) are similar over the years, which can be associated with the minimal change in the source term of alpha activity concentrations during this five-year period. The averages are always higher than P50. The average ranges from 60.2 µBq/m^3 in 2018 to 72.9 µBq/m^3 in 2015, and the 90th percentile (P90), which is usually taken as reference of maximum values [19], to avoid the impact of possible outliers, ranges from 108.1 μBq/m^3 in 2014 to 117.2 μBq/m^3 in 2015.

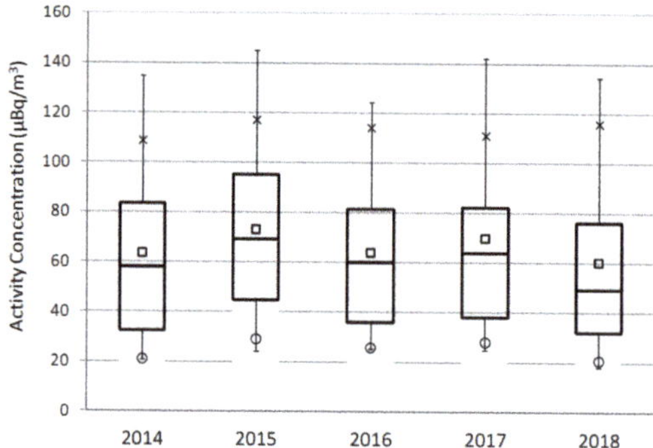

Figure 5. Box-and-whisker plot of alpha activity concentrations over the years (2014–2018) at Bilbao. The meaning of the box plot is the same of Figure 3b.

The analysis of the intra-annual variability, considering winter (December, January, and February), spring (March, April, and May), summer (June, July, and August), and Autumn (September, October, and November), reports the maximum average value in autumn (Figure 6a), concretely in October (110 μBq/m^3) (Figure 6b), while the minimum average value is in winter (Figure 6a), with the monthly minimum in February (40 μBq/m^3) (Figure 6b). These seasonal and monthly variability do not agree with previous studies carried out in Spain [5,8,9], in which the activity maxima are registered in summer. This difference would point out differences in source terms and in the meteorological conditions driven alpha activity concentrations during the years among these sites, and hence, the need to analyze the link between alpha and meteorological parameters.

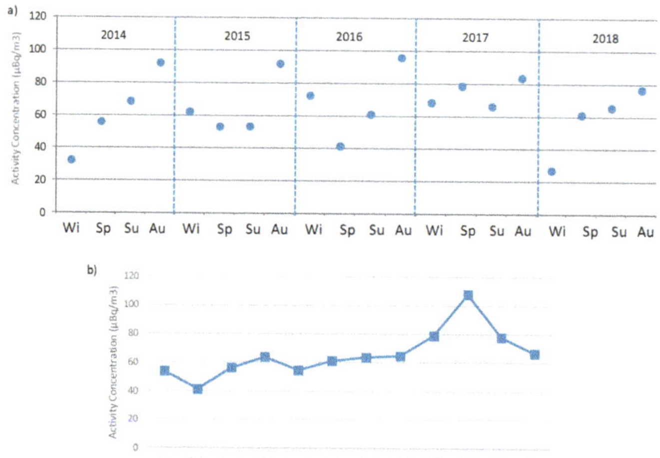

Figure 6. (**a**) Stational variability and (**b**) monthly average of alpha activity concentrations at Bilbao from 2014 to 2018.

3.2. Meteorological Characterization of the Alpha Peak Activity Concentrations

This section presents the results of studying the impact of meteorological parameters on the highest alpha activity concentrations registered in Bilbao for the five years analyzed (2014–2018).

The reason for taking the highest alpha activity concentrations as reference is due to health hazards associated with the presence of radionuclides in air [20].

High alpha surface concentrations are identified as values exceeding the P90 calculated over 2014–2018 (117.1 µBq/m3 in Figure 3b). P90 has been previously used in other research as the extreme criterion [18–21]. The average of the 26 high alpha values identified during the period 2014–2018 above P90 is 152.0 ± 6.6 µBq/m^3, while the highest alpha value is 246.5 µBq/m^3. Seasonal distribution of these alpha peak concentrations is shown in Figure 7, as well as the corresponding seasonal mean. This Figure 7 reports the largest occurrence of these high alpha concentrations in autumn (14 out of 26 periods, 53%), and the minimum occurrence in summer (three out of 26 periods, 11.5%). However, the maximum seasonal average of these peak concentrations is registered in spring (187.9 µBq/m^3), followed by summer (166.7 µBq/m^3), and lower and similar values in winter (142.4 µBq/m^3) and autumn (144.8 µBq/m^3). This seasonal variability of alpha peak concentrations, mainly explained by differences in atmospheric factors, is different to Figure 6a but similar to the seasonal variations shown in previous articles carried out in Spain, and, at the same time, it provides insight into the large impact of occasional high alpha values during the warm (spring/summer) period of the year.

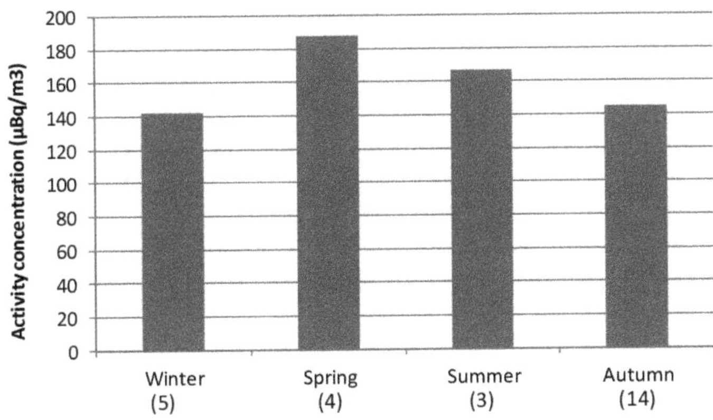

Figure 7. Averaged concentration in each season considering only those sampling periods registering alpha concentrations above P90 during 2014–2018 (in brackets below the season).

The evaluation of correlation between weekly alpha peak activity concentrations and arithmetic values of local meteorological parameters (temperature, relative humidity, atmospheric pressure, and wind speed and direction) during each sampling period is performed at the significance level of 0.05. In the case of precipitation, the total amount during each sampling period is used. The correlations found are around zero, i.e., there is no relationship between the two variables used. The highest is −0.22 with precipitation, which, being negative, indicates that precipitation tends to reduce alpha activity at the site as a consequence of scavenging of airborne radionuclides by rainfall [22]. These results, with none of them significant at 0.05, indicate the weakness of the relationship between alpha peak concentrations and each individual meteorological variable at this weekly temporal scale level. Therefore, other atmospheric processes should be included in the present analysis aiming at understanding the meteorological reasons of these high alpha surface concentrations in Bilbao.

Due to the impact that airflow patterns have over the surface weather conditions and aerosols temporal distribution in a given area and over a period of time [23], backward trajectories in Bilbao are calculated and are used to determine the influence of regional meteorological patterns on alpha peak concentrations. Figure 8a shows the mean trajectories (cluster) for Bilbao at 100 m calculated considering all the 26 sampling periods above P90. Seven airflow patterns are identified by applying the cluster methodology explained in Stunder [15]. Taking into account the pathways followed by each airflow pattern as well as the frequency (number in brackets) of each one, the arrival of slow continental southern (27%, cluster 4) and maritime northern (28%, cluster 1) air masses dominates periods with high alpha concentrations against the arrival of fast northern (7%, cluster 6) and different Atlantic advections from the northwest (24%, clusters 2 and 3) and from the west (14%, clusters 5 and 7) (Figure 8a).

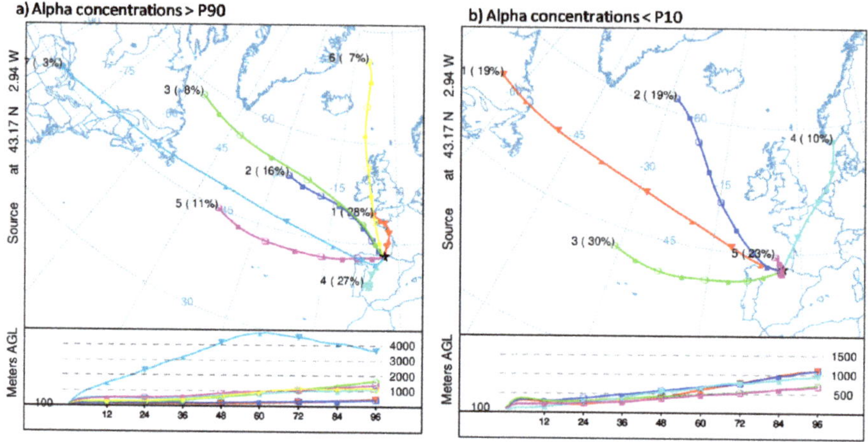

Figure 8. Back-trajectory cluster centers (centroids) obtained at 100 m agl for periods with alpha activity concentrations (**a**) above the 90th percentile and (**b**) below the 10th percentile during the 2014–2018 period at Bilbao. The left numbers in the centroids are an identification number of the centroid and the right numbers (in brackets) are the percentage of complete trajectories occurring in that cluster.

To establish the link of these airflow patterns with alpha peak concentrations, we have also calculated the airflow patterns associated with the lowest alpha concentrations registered in Bilbao (26 periods in which alpha concentrations were below the 10th percentile) (Figure 8b). The comparison of both (Figure 8) points out similarities between airflow patterns, but with notable differences in frequencies (number in brackets) in those similar. The lowest alpha concentrations highlights the higher occurrence of western airflows, and air masses from the south and north, with slow displacement (Cluster 1 and Cluster 4 in Figure 8a), are not identified. Therefore, both Figures mainly suggest that the arrival of continental southern and maritime northern airflows is mainly associated with measurements of high alpha concentrations in this area.

To study the relationships between airflow patterns (Figure 8a) and alpha peak concentrations, and with the purpose to not consider the same sampling period within two different airflow patterns, which would create misunderstanding in the results, we have identified persistent sampling periods, i.e., periods in which most of the trajectories (above 75% of the calculated trajectories in each period) are grouped into the same cluster. Ten out of 26 sampling periods are identified as persistent ones: six within cluster 1, one within cluster 2, and three within cluster 4. Figure 9 shows the summary of alpha peak concentration measured and the seasonal distribution of these persistent sampling periods. They occurred in autumn/winter seasons in cluster 1 and cluster 4, while the only persistent period corresponding to cluster 2 took place in spring. This result, hence, reports that sampling periods characterized by prevailing southern and northern airflows seems to drive the occurrence of alpha peak activity concentrations in this area. The fact of mainly being located in autumn/winter, which is the wettest period in Bilbao (see Section 2.1), would not be in line with previous studies [6], where alpha activity concentrations decrease with precipitation. However, and due to the geographic location of the region, winds coming from the south descending from the Cantabrian Mountains became warm and dry because the Foehn effect [24], and even within the wettest period, under the arrival of southerly winds, the chances of rain are nearly zero in the region even.

	Number of persistent periods	Mean (µBq/m³)	SD	Max (µBq/m³)	Min (µBq/m³)
Cluster 1	6	159	19	247	122
Cluster 2	1	127			
Cluster 4	3	152	2	157	149

Figure 9. Statistical values in air for alpha peak concentrations during the persistent sampling periods at Bilbao and the corresponding seasonal distribution of the periods. The uncertainties of the means are indicated as the standard deviation of the average Sx/N1/2, where N is the number of persistent periods and the standard deviation. Clusters refer to Figure 8a.

3.3. Case Studies

Case studies are needed to evaluate on a real basis the potential influence of each airflow pattern and the associated meteorological conditions on the alpha peak concentrations in Bilbao. So, we have analyzed representative periods of the two main meteorological scenarios driving these concentrations, i.e., the arrival of continental southern and maritime northern air masses. To analyze the influence of each airflow pattern on alpha activity concentrations, Figure 10 displays surface winds (blowing from) during one persistent sampling period (02–08/10/2018) under northerly flows, and from 9 December 2015 to 14 December 2015 under southerly flows. Hourly evolution of ^{222}Rn is also shown in both periods to represent the impact of surface winds on activity concentrations.

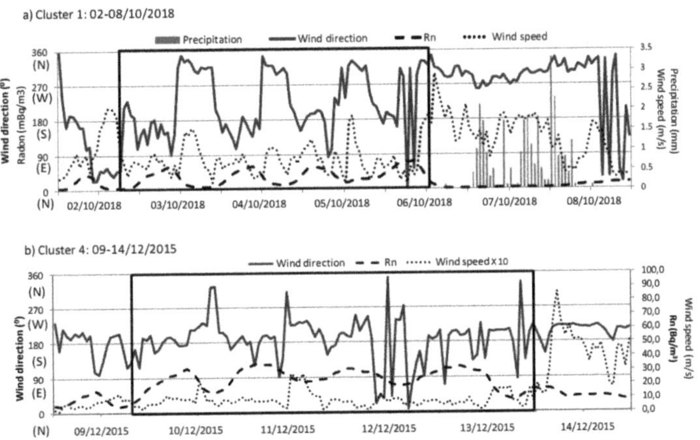

Figure 10. Daily variation of 222Rn activity concentration, wind speed and direction, and total amount of precipitation at Bilbao between (**a**) 2 October 2018 and 8 October 2018 (northerly flows-cluster 1) and (**b**) 9 December 2015 and 14 December 2015 (southerly flows-cluster 4). Note the differences in scales.

The alpha activity concentration recorded in the first sampling period is 122.5 µBq/m^3. According to the wind direction variability, two different periods can be identified in these seven days of sampling (Figure 10a). This fact allows considering this sampling period as suitable to analyze differences in surface winds under the same synoptic meteorological scenario, and its relationship with surface activity concentrations. In addition, at the end of this sampling period was constantly raining.

The first four days, the synoptic conditions favored the progressive development of a daily evolution cycle of surface winds, blowing from the south during the night, followed by the clockwise swing to northwesterly winds during the day. Wind speed was less than 2 m/s. This daily evolution of winds at surface level agrees with the development of sea/land breezes in this area [25], which are more common during spring and summer. Under this surface condition, and considering the marked daily cycle of ^{222}Rn [26], it is suggested an accumulation process, which progressively raises the concentration of particles close to the surface in this area, i.e., it traps the particles near the surface and limits a sufficient mixing with the air above [27]. The limitation of the dispersion processes would thus favor the increase in the alpha activity concentration in this area. This daily cycle is broken at the end of the sampling period, with a sudden change in surface winds, which started blowing constantly from the northwest-north range and with wind speeds higher than in the previous days. Under these new conditions, the dispersion processes are favored, and together with the sporadic rainfall during these two days, cause a decrease in activity concentrations in the area, as it well observed in the daily evolution of ^{222}Rn concentrations.

The alpha activity concentration recorded in the second sampling period is 149.6 μBq/m^3, and there is no rainfall measured in these six days. The synoptic meteorological conditions allowed the constant arrival of surface winds from the south. In this period, changes in wind speed allow identifying two different surface dynamics. In the first five days, the southerly pattern is developed with similar and low wind speeds, which favored similar daily activity concentrations but limited the accumulation processes, not registering high ^{222}Rn activity concentrations (comparing with Figure 10a, the radon concentrations differ by about two orders of magnitude). In the last day, the increase in wind speed cause a cleaning effect by enhancing the dispersive effect in the surface atmospheric layers, as can be seen in Figure 10b following the evolution of ^{222}Rn activity concentrations.

4. Conclusions

Alpha activity concentration can be usually used as a control indicator of the radioactivity levels in air, and so a low-flow station is located in Bilbao, northern Spain. Here, we have analyzed 258 data registered from 2014 to 2018. These alpha activity concentration data can be considered as a time series after autocorrelation function confirms its behavior and periodogram shows an annual periodicity and a significant seasonal and monthly variation of alpha activity concentrations.

In order to explain the alpha peak concentrations, the influence of the airflow patterns and local meteorological parameters has been studied. Precipitation presented a high negative correlation with alpha values, indicating removal process of airborne and the decrease in concentrations. Slow air masses from the south (continental displacement) and from the north (Atlantic displacement) have been found as mainly responsible for the alpha peak concentrations in this area. Under the arrival of northern airflows, the wind directions at the surface levels displayed the development of sea-land breezes, with northern and southern winds, while under the arrival of continental airflows, the arrival of southern winds at surface levels dominates. Analyzing the hourly evolution of ^{222}Rn together with surface winds, we have identified how the first meteorological scenario favors the accumulation processes and hence drastically increases activities, while the second scenario keeps activity concentrations similar but with lower values than in the previous scenario during the sampling period.

Author Contributions: Conceptualization, M.Á.H.-C. and N.A.; methodology, M.Á.H.-C. and N.A.; software, M.Á.H.-C. and N.A.; formal analysis, M.Á.H.-C. and N.A.; writing—original draft preparation, M.Á.H.-C. and N.A.; writing—review and editing, M.Á.H., N.A. and F.L. All authors have read and agreed to the published version of the manuscript.

Funding: This research received no external funding.

Acknowledgments: The authors gratefully acknowledge the NOAA Air Resources Laboratory (ARL) for the provision of the HYSPLIT transport and dispersion model and/or READY website (https://www.ready.noaa.gov) used in this publication.

Conflicts of Interest: The authors declare no conflict of interest.

References

1. Bossew, P.; Cinelli, C.; Hernández-Ceballos, M.A.; Cernohlawek, N.; Gruber, V.; Dehandschutter, B.; Menneson, F.; Bleher, M.; Stöhlker, U.; Hellmann, I.; et al. Estimating the terrestrial gamma dose rate by decomposition of the ambient dose equivalent rate. *J. Env. Radioact.* **2017**, *166*, 296–308. [CrossRef] [PubMed]

2. EPA. Radiological Laboratory Sample Analysis Guide for Incidents of National Significance—Radionuclides in Air. Available online: https://www.epa.gov/sites/production/files/2015--05/documents/402-r-09--007-air-guide.pdf (accessed on 27 October 2020).

3. Papastefanou, C. Radioactive aerosols. *Radioact. Environ.* **2008**, *12*, 11–58.

4. Tositti, L.; Brattich, E.; Cinelli, G.; Baldacci, D. 12 years of 7Be and 210Pb in Mt. Cimone, and their correlation with meteorological parameters. *Atmos. Env.* **2014**, *87*, 108–122. [CrossRef]

5. Dueñas, C.; Orza, J.A.; Cabello, M.; Fernández, M.C.; Cañete, S.; Pérez, M.; Gordo, E. Air mass origin and its influence on radionuclide activities (7Be and 210Pb) in aerosol particles at a coastal site in the western Mediterranean. *Atmos. Res.* **2011**, *101*, 205–214. [CrossRef]

6. Dueñas, C.; Fernández, M.C.; Cañete, S.; Pérez, M. [7]Be to ^{210}Pb concentration ratio in ground level air in Málaga (36.7°N, 4.5°W). *Atmos. Res.* **2009**, *92*, 49–57.

7. Piñero-García, F.; Ferro-García, M.A.; Chham, E.; Cobos-Díaz, M.; González-Rodelas, A.P. A cluster analysis of back trajectories to study the behaviour of radioactive aerosols in the south-east of Spain. *J. Environ. Radioact.* **2015**, *147*, 142–152. [CrossRef] [PubMed]

8. García-Talavera, M.; Quintana, B.; García-Díez, E.; Fernández, F. Studies on radioactivity in aerosols as a function of meteorological variables in Salamanca (Spain). *Atmos. Env.* **2001**, *35*, 221–229. [CrossRef]

9. Sáez-Muñoz, M.; Bas Cerdá, M.C.; Ortiz, J.; Martorell, S. Analysis of the evolution of gross alpha and gross beta activities in airborne samples in Valencia (Spain). *J. Env. Radioact* **2018**, *183*, 94–101. [CrossRef]

10. Johnson, G.R.; Morawska, L. The mechanism of breath aerosol formation. *J. Aerosol Med. Pulm. Drug Deliv.* **2009**, *22*, 229–237. [CrossRef]

11. Charles, M. Sources and effects of ionizing radiation. *J. Radiol. Prot.* **2000**, *21*, 1.

12. European Atlas of Natural Radiation. Available online: https://analytik.news/en/press/2020/148.html (accessed on 27 July 2020).

13. Stein, A.F.; Draxler, R.R.; Rolph, G.D.; Stunder, B.J.B.; Cohen, M.D.; Ngan, F. NOAA's HYSPLIT atmospheric transport and dispersion modeling system. *Bull. Amer. Meteor. Soc.* **2015**, *96*, 2059–2077. [CrossRef]

14. Kassomenos, P.; Vardoulakis, S.; Borge, R.; Lumbreras, J.; Papaloukas, C.; Karakitsios, S. Comparison of statistical clustering techniques for the classification of modelled atmospheric trajectories. *Appl. Clim.* **2010**, *102*, 1–12. [CrossRef]

15. Stunder, B. An assessment of the quality of forecast trajectories. *J. Appl. Meteorol.* **1996**, *35*, 1319–1331. [CrossRef]

16. Kendall, M.; Ord, J.K. *Time Series*, 3rd ed.; Charles Griffin: London, UK, 1976; p. 296.

17. Hewitt, C.N. *Methods of Environmental Data Analysis*; Elsevier Applied Science: London, UK, 1992; 343p.

18. Chham, E.; Milena-Pérez, A.; Piñero-García, F.; Hernández-Ceballos, M.A.; Orza, J.A.G.; Brattich, E.; El Bardouni, T.; Ferro-García, M.A. Sources of the seasonal-trend behaviour and periodicity modulation of 7 Be air concentration in the atmospheric surface layer observed in southeastern Spain. *Atmos. Environ.* **2019**, *213*, 148–158. [CrossRef]

19. Lozano, R.; Hernández-Ceballos, M.A.; San Miguel, E.G.; Adame, J.A.; Bolívar, J.P. Meteorological factors influencing the [7]Be and ^{210}Pb concentrations in surface air from the southwestern Iberian Peninsula. *Atmos. Environ.* **2012**, *63*, 168–178. [CrossRef]

20. UNSCEAR. Sources and Effects of Ionizing Radiation. Available online: https://www.scirp.org/reference/referencespapers.aspx?referenceid=2735033 (accessed on 27 October 2020).

21. San Miguel, E.G.; Hernández-Ceballos, M.A.; García-Mozo, H.; Bolívar, J.P. Evidences of different meteorological patterns governing 7 Be and 210Pb surface levels in the southern Iberian Peninsula. *J. Environ. Radioact.* **2019**, *198*, 1–10. [CrossRef]

22. Zheng, M.; Sjolte, J.; Adolphi, F.; Aldahan, A.; Possnert, G.; Wu, M. Muscheler, R. Solar and meteorological influences on seasonal atmospheric 7Be in Europe for 1975 to 2018. *Chemosphere* **2020**, *263*, 128318. [CrossRef]

23. Pirouzmand, A.; Kowsar, A.; Dehghani, P. Atmospheric dispersion assessment of radioactive materials during severe accident conditions for Bushehr nuclear power plant using HYSPLIT code. *Prog. Nucl. Energy* **2018**, *108*, 169–178. [CrossRef]

24. Usabiaga, J.I.; Sáenz, J.; Valencia, V.; Borja, A. Chapter 4 - Climate and Meteorology: Variability and its influence on the Ocean. *Elsevier Oceanogr. Ser.* **2004**, *70*, 75–95.

25. Acero, J.A.; Arrizabalaga, J.; Kupski, S.; Katzschner, L. Urban heat island in a coastal urban area in northern Spain. *Theor. Appl. Climatol.* **2013**, *113*, 137–154. [CrossRef]

26. Chham, E.; Piñero-García, F.; González-Rodelas, P.; Ferro-García, M.A. Impact of air masses on the distribution of 210Pb in the southeast of Iberian Peninsula air. *J. Environ. Radioact.* **2017**, *177*, 169–183. [CrossRef]

27. Garratt, J.R. The Atmospheric Boundary Layer. *Earth-Sci. Rev.* **1994**, *37*, 89–134. [CrossRef]

International Journal of
*Environmental Research
and Public Health*

Article

Role of Meteorological Parameters in the Diurnal and Seasonal Variation of NO$_2$ in a Romanian Urban Environment

Mirela Voiculescu, Daniel-Eduard Constantin *, Simona Condurache-Bota,
Valentina Călmuc, Adrian Roșu and Carmelia Mariana Dragomir Bălănică

Faculty of Sciences and Environment, European Center of Excellence for the Environment, "Dunărea de Jos"
University of Galați, 800008 Galați, Romania; Mirela.Voiculescu@ugal.ro (M.V.); scondurache@ugal.ro (S.C.-B.);
valentina.calmuc@ugal.ro (V.C.); adrian.rosu@ugal.ro (A.R.); carmelia.dragomir@ugal.ro (C.M.D.B.)
* Correspondence: Daniel.Constantin@ugal.ro

Received: 16 July 2020; Accepted: 24 August 2020; Published: 27 August 2020

Abstract: The main purpose of this study was to investigate whether meteorological parameters (temperature, relative humidity, direct radiation) play an important role in modifying the NO$_2$ concentration in an urban environment. The diurnal and seasonal variation recorded at a NO$_2$ traffic station was analyzed, based on data collected in situ in a Romanian city, Braila (45.26° N, 27.95° E), during 2009–2014. The NO$_2$ atmospheric content close to the ground had, in general, a summer minimum and a late autumn/winter maximum for most years. Two diurnal peaks were observed, regardless of the season, which were more evident during cold months. Traffic is an important contributor to the NO$_2$ atmospheric pollution during daytime hours. The variability of in situ measurements of NO$_2$ concentration compared relatively well with space-based observations of the NO$_2$ vertical column by the Ozone Monitoring Instrument (OMI) satellite for most of the period under scrutiny. Data for daytime and nighttime (when the traffic is reduced) were analyzed separately, in the attempt to isolate meteorological effects. Meteorological parameters are not fully independent and we used partial correlation analysis to check whether the relationships with one parameter may be induced by another. The correlation between NO$_2$ and temperature was not coherent. Relative humidity and solar radiation seemed to play a role in shaping the NO$_2$ concentration, regardless of the time of day, and these relationships were only partially interconnected.

Keywords: nitrogen dioxide; in situ urban concentrations; meteorological measurements; NO$_2$ variation; partial correlation

---

1. Introduction

Monitoring atmospheric pollution and the possibility to predict its evolution are of high interest. One major atmospheric pollutant is nitrogen dioxide (NO$_2$), which may cause many health problems [1]. The NO$_2$ gas has a reddish-brown color, is nonflammable, and has a detectable, pungent odor, perceptible from concentrations of approximately 190 µg/m^3 [2]. Nitrogen dioxide reacts with the hydroxyl radical (-OH) in the atmosphere, forming the highly-corrosive nitric acid, but it can also form toxic organic nitrates. Nitrogen oxides known as NO$_x$ (i.e., NO + NO$_2$) are involved in the formation of tropospheric ozone and smog, mediated by light through photolysis. Due to the significant environmental impact, the NO$_x$ compounds are strictly monitored and legislative limits were set at EU and national levels [2]. Part of the NO$_2$ molecules in the atmosphere are of primary nature (i.e., directly emitted), while most NO$_2$ results from nitrogen monoxide (NO) [1,3]. The latter is produced via natural and anthropogenic processes. About one-fifth of NO$_x$ is released in the atmosphere during thunderstorms, through lightning, in the upper troposphere [3]. Other natural NO_x sources include volcanic activity emissions, bacteria decay, soil processes and stratospheric sources [1], while the rest is the result of fossil fuel combustion processes: fires, transportation, industrial—such as power generation, nitric acid manufacture, welding processes, the use of explosives and iron and steel industry and refineries [1,2]. According to [4], about 65% of NO_x emissions are of anthropogenic origin. The NO_2 reaction with the atmospheric hydroxyl radical (-OH), and the NO reaction with hydroperoxyl (HO_2), followed by the formation of nitric acid and its wet deposition are important NO sinks (almost 60%), according to [4]. The contribution of road traffic to nitrogen oxides (NO_x) emissions has been described by various studies [5–11]. Some authors assume that about half of the anthropogenic NO_x comes from traffic [12]. Although several measures of emission control were implemented by various European Commission (EC) regulations, emissions of nitrogen oxides from road traffic have not significantly decreased across Europe; sometimes, NO_2 concentrations have even increased [10,11]. Some recent papers showed that the real reduction of nitrogen oxides and other pollutants following Euro 5 and 6a/b introduction was not as important as expected [10].

The concentration of NO_2 (and, in general, of any atmospheric pollutants) in an urban area and its variation are influenced by the industrialization level, the number of inhabitants and traffic density, but also by the topography of the area and meteorological parameters [5,9,11]. It is expected that meteorological factors would affect both natural and anthropogenic emissions of NO_2 (e.g., the reduced solar radiation during wintertime will impact the rate of NO_2 photochemical reactions and the reduced temperatures during the cold season will increase NO_2 emissions associated with heating). Moreover, meteorological parameters are interlinked in various ways: the relative humidity decreases with increasing temperature, the effective and saturation air vapor pressures increase with atmospheric temperature, etc. Thus, it is difficult to assess the effect of each separate meteorological factor on NO_2 pollution. None of these influences were yet quantified in a reliable way and a clear relationship is far from being established. Studies of the link between meteorological factors and atmospheric pollutants, in general, or NO_2, in particular, are rather scarce, inconsistent and strongly depend on local factors [5,13–19]. According to [13], the NO_2 concentration is slightly higher at a lower relative humidity, whereas other authors found that the NO_2 concentration correlates positively with the relative humidity in all seasons, especially during winter [20]. The dependence between NO_2 and temperature was found to be weak, in general, except for two considerable positive correlations, which were obtained for July and December, and for which no particular explanation was found in [13]. Other studies found strong positive correlations between the NO_2 concentration and temperature for all seasons [20]. A negative correlation between wind speed and NO_2 concentration in all seasons except for summer was found by [19].

This paper presents a study of the diurnal, monthly, seasonal and interannual evolution of nitrogen dioxide concentration at an urban site in the southeast of Romania (Braila, 45.3° N, 27.9° E), which is influenced by road traffic mainly during the daytime. The role played by anthropogenic sources and the most relevant meteorological parameters that may affect the level of NO_2 pollution at the local scale is investigated. A comparison between in situ measurements of the NO_2 concentration near the ground and satellite measurements of NO_2 vertical columns is also performed. The timeframe for both in situ and satellite data is 2009–2014.

2. Data and Methods/Experimental

2.1. Data

The experimental data used in this study are from the automated monitoring air quality station called "BRAILA 1", denoted as "BR1", which is a traffic type station, one of the four available stations in Braila town, which are part of the National Network for Monitoring Air Quality (RNMCA, The Agency of Environmental Protection Braila: http://apmbr.anpm.ro/). The county of Braila, situated on the right shore of the Danube River, is considered to be the 11th Romanian city by the number of inhabitants.

The number of vehicles in Braila has increased during the last few years, which led to a rather increased concentration of atmospheric pollutants. BR1 is a traffic station; it is located on Calea Galați No. 53 (The Agency of Environmental Protection Braila: http://apmbr.anpm.ro/) and monitors on a continuous basis the pollution levels generated mainly by traffic emissions, with medium and high flows, from the neighboring streets. Calea Galati Street is one of the busiest traffic streets in Braila. The width of the street is 16 m and it is covered with an asphalt carpet. The area around the air quality monitoring station has apartment blocks with four floors and some green spaces. The air quality monitoring station is located approximately 20 m from Calea Galati Street and 15 m from the apartment buildings. The surface is flat, characteristic of a plain area. The sources of pollution in the area consist mostly of road traffic and domestic heating.

The hourly atmospheric concentrations of NO_2 are determined in situ, by using the chemiluminescence technique [8,21]. This method implies the reduction of NO_2 to NO in the presence of a Molybdenum surface, at a temperature of roughly 310 °C. The resulted NO reacts with ozone (O_3), leading to the formation of fluorescent NO_2, whose emission is detected by a sensor [4,21]. Data between 2009 and 2014 were used in this study. Besides, concentrations of NO_2, meteorological parameters were also recorded.

Satellite measurements were provided by the Ozone Monitoring Instrument (OMI) onboard the Earth Observing System Aura satellite [22,23]. The OMI measures several important pollutants, such as O_3, NO_2, SO_2, and aerosols, with a spatial resolution of 13 km × 24 km, i.e., at a near urban scale resolution, https://aura.gsfc.nasa.gov/omi.html [24]. OMI is a remote sensing space instrument onboard the Aura satellite that provides a raw NO_2 product called DSCD (Differential Slant Column Density), which is retrieved using Differential Optical Absorption Spectroscopy (DOAS) in the 405–465 nm range. Satellite measurements provide valuable data about atmospheric pollutants, including NO_2 [25–32] at a large scale. The most refined product of OMI is the tropospheric Vertical Column Density (VCD), which is the result of a near real-time retrieval algorithm that gives a 0.7×10^{15} molec./cm^2 uncertainty for each individual pixel [26]. In our study, we have used level 3 data, which are a conversion of OMI 13 × 24 km^2 data to a 0.25° × 0.25° resolution [33]. Such data collected between 2005 and 2015 by OMI confirmed a reduction in NO_2 concentration over Northern China, Eastern Europe, and USA, while an increase in NO_2 concentration was detected over the Persian Gulf and India [30].

2.2. Methods

Meteorological parameters recorded at an hourly rate were used, in order to verify to what extent they are relevant for modulating the NO_x variability by means of correlation analysis. Correlations/anticorrelations (i.e., positive/negative correlation coefficients) between two variables are not the decisive proof for direct cause–effect relationships; however, they suggest that a link may exist if a physical mechanism supports it. When two variables are not correlated, the first thought is that there is no link between the two variables, which, however, may not always be true. Moreover, meteorological parameters (temperature, humidity, solar radiation) are not independent (e.g., solar radiation at the ground is a measure of the cloud cover, which is linked to relative humidity but also to temperature). Consequently, the correlation analysis was refined and a partial correlation analysis was used [34,35], complementary to the bivariate correlation. Such an analysis is useful when the relationship between two variables, e.g., X and Z, measured by the bivariate correlation coefficient, C_{XZ}, may be induced or suppressed by a third (intervening) variable, e.g., Y, which affects both variables X and Z. The partial correlation $P_{X(Y)Z}$ corresponds to the link between X and Z when Y is constant. To be more specific, in this particular case the main variables, X and Z, were the NO_2 concentration and one meteorological factor (e.g., temperature) and the intervening variable, Y, was another meteorological parameter (e.g., radiation). The radiation (Y) may affect both the NO_2 concentration (X) and the temperature (Z). The difference $D_{X(Y)Z} = P_{X(Y)Z} - C_{XZ}$ between the partial correlation coefficient, $P_{X(Y)Z}$, and the direct correlation coefficient, C_{XZ}, measures the degree of intervention for Y. If $P_{X(Y)Z}$ is smaller than C_{XZ}, i.e., the difference and the direct coefficient have opposite signs and $C_{XZ} \times D_{X(Y)Z} < 0$, then Y is responsible for part of the correlation, or it is said that Y "intervenes". If the two coefficients, C and D, have the same sign, i.e., $C_{XZ} \times D_{X(Y)Z} > 0$, then the correlation is real and Y even suppresses, partially, the correlation. If the partial and direct coefficients are equal, i.e., $D = 0$, then the Y variable does not intervene. Table 1 summarizes possible results and interpretations for various combinations of direct and partial correlation coefficients.

Table 1. Type of correlation for various results of the partial correlation analysis.

C (Bivariate/Direct Correlation)	$D = P - C$	$C \times D$	Correlation between X and Z	Role of the Intervening Variable (Y)
>0	0	0	real (positive) correlation	none
>0	<0	<0	partially spurious	partially responsible for the correlation
>0	>0	>0	partially suppressed	partially masks the real correlation
0	any	0	fully suppressed	fully masks the real correlation
<0	0	0	real (negative) correlation	none
<0	<0	>0	partially suppressed	partially masks the real correlation
<0	>0	<0	partially spurious	partially responsible for the correlation
any	=C	>0	fully spurious	fully responsible for the correlation

The correlation analysis has been applied to standardized data (by their standard deviation) in order to identify possible links between NO_2 concentration in the atmosphere and the variations of several meteorological parameters.

3. Results and Discussion

3.1. Diurnal and Seasonal Variation

Figure 1 shows the diurnal and seasonal variation of NO_2 concentration (top) and temperature (bottom) during 2009–2014. The hourly averaged NO_2 concentration varied between 14 $\mu g/m^3$ (July 2009) and 41 $\mu g/m^3$ (January 2011). These values are smaller than those measured by traffic stations in Bucharest [28].

The NO_2 content varied with year; it was low in 2009 and in the first part of 2010, reached a maximum in the winter between 2010 and 2011, and then slightly decreased towards the end of the interval. The period is too short for assessing whether a trend does exist.

The seasonal variation of NO_2 was evident; NO_2 concentrations were clearly higher during winter than during summer. The seasonal variation may be explained by: (1) the strong variation of the anthropogenic sources contributing to NO_2 emissions, i.e., traffic and combustion, but also by (2) a longer lifetime of the NO_2 during winter, taking into account that the NO_2 lifetime and its concentration in the atmosphere are affected by the seasonal variation of the photochemical activity [7,8,26,31], which is reduced during winter. Additionally, during winter, the soil is cold, and thus the nearby air is heavy so that the emitted pollutants may stay close to the ground longer [2]. Another explanation for the seasonal NO_x variation relates to the variation of the boundary layer height with the season, which influences the wind pattern that, in turn, has a very important role in pollutant dispersion [7].

Two diurnal peaks of the NO_2 concentration could be observed, centered around 09:00 Local Time (LT) and 20:00 LT, which were more evident during cold months. The diurnal peaks are in agreement with previous findings [8] and are associated with peaks in traffic [7]. Morning peaks, observed around 9:00, were smaller than the evening peaks. The NO_2 concentration increased during the afternoon-evening period, between 17:00 and 24:00, with maxima progressing towards earlier time during winter months. This was seen in all seasons except summer. In December, these diurnal maxima were smaller, for all years, which can be linked to the reduced traffic and industrial activity due to holidays.

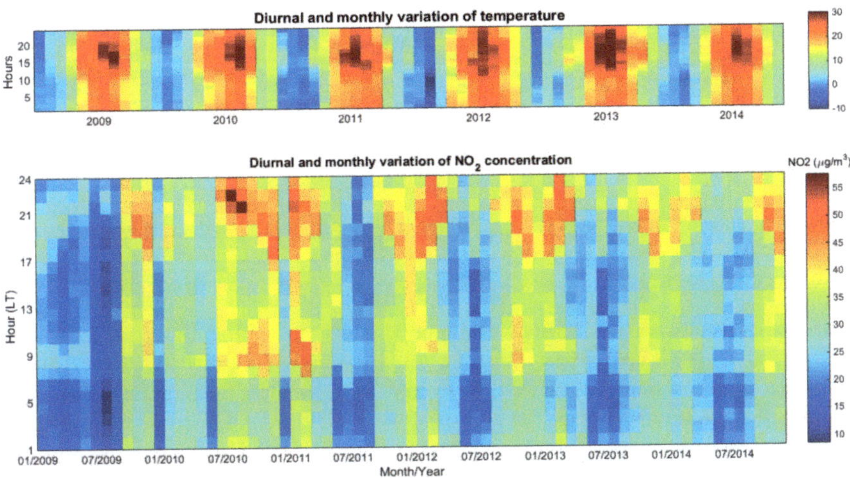

Figure 1. Diurnal and seasonal variation of NO_2 concentration (**bottom**) and of temperature (**top**) during 2009–2014 (monthly means).

Figure 2 shows the variation of the tropospheric NO_2 VCD (Vertical Column Density) over Braila, derived from OMI, together with in situ measurements at 11:00 LT, which is the approximate hour when the satellite overpasses the city. The tropospheric NO_2 VCD varied between 0.8×10^{15} molec./cm^2 for February 2011 and 4.7×10^{15} molec./cm^2 for December 2012. Note that the comparison in Figure 2 is strictly qualitative, since OMI measures the number of NO_2 molecules in a column over a grid of 0.25° [33], while in situ instruments measure the volumetric concentration near the ground; thus a direct quantitative comparison makes no sense. The annual NO_2 VCD ranges between 1.73 and 2.23×10^{15} molec./cm^2. Previous studies have shown that OMI measurements give reliable results about NO_2 emissions of anthropogenic sources at a large scale in cities [26,32,36], above large industrial platforms (located away from cities) [25,37] and given by vehicle emissions and also by residential emissions [38]. However, other studies have shown that satellite observations do not correctly evaluate the NO_2 pollution caused by traffic or by other point sources [27,28].

Figure 2 shows that the two time series agreed relatively well except for a period between the winter of 2009–2010 and the summer of 2011. The correlation coefficient between the two series was close to 0.40 (after values in December 2010 were removed) and did not change when NO_2 concentrations measured earlier, between 08:00 and 11:00 LT, were considered for the mean.

The seasonal variation of the tropospheric VCD was more regular: it was small during warm months and high during cold months. In 2010, the two time series did not coincide, which is mainly due to a peculiar behavior of the in situ measurements that showed an unexpected wide maximum during summer and autumn, and a minimum in December. The OMI instrument "sees" a large surface area (312 km^2), and thus integrates the emissions of many sources. Consequently, it cannot accurately measure the local variability, captured by the in situ measurements. This opposing variation suggests that a punctual, temporary, source of NO_2 was added in the second part of 2010 close to the measuring station.

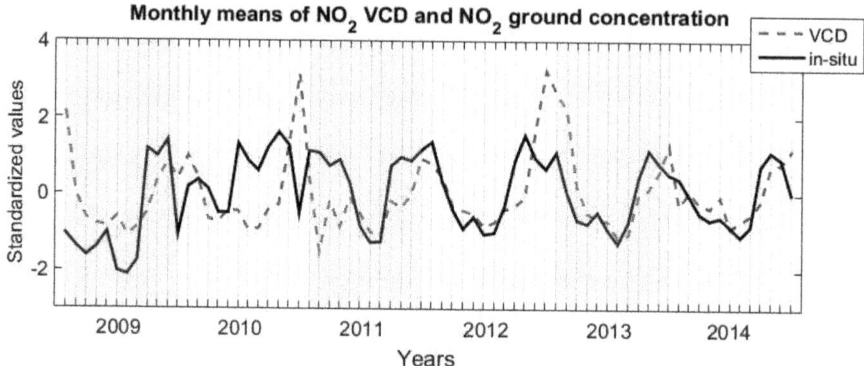

Figure 2. Variation of standardized monthly averages of tropospheric NO_2 (VCD measured by OMI, blue, dash) and of NO_2 ground concentration measured in situ at 11:00 LT (black, solid) starting with January 2009 until December 2014.

Figure 3 shows the average diurnal variation of the NO_2 concentration during each season for the entire period. Seasons are defined here in a slightly different manner than usual: spring and fall/autumn consist of two months each (March–April and September–October), while solstice seasons have four months: the summer covers the period between May and August, and the winter starts in November and ends in February. This definition of seasons is justified by the weather profile during the last 20 years in Braila County and in South-Eastern Romania, which shows a fast transition from winter to summer and then back to winter.

In general, the diurnal variation was relatively regular for fall, when a relatively small data spread was seen. A higher variability was observed in winter, spring and summer, when the data spread was larger. There were no outliers in the middle part of the day, when the traffic is slightly reduced compared to the morning and afternoon. The outliers during morning and afternoon may be associated with peaks or drop-offs of the traffic. The least regular season was summer, when the number of outliers was high regardless of the hour and their large majority lied in the upper part of the plot. This suggests that the NO_2 loading was significantly higher than the average for a short period of time. This is confirmed by the next analysis (Figure 4). The explanation might relate to road construction works that resulted in deviation of the traffic. Starting with 2010, landslides occurred on streets near the measuring station. In 2011, the asphalt was removed and some excavations were done, in order to check the underground water and sewerage pipes. Subsequently, utility pipes were installed, the foundation was compacted and the traffic flow returned to normal values.

Apparently, the NO_2 concentration and the temperature were anticorrelated, especially during daytime: high NO_2 concentrations during morning corresponded to the lowest temperatures, while the afternoon reduction in NO_2 coincided with the highest temperatures. However, this is just coincidental, since the daytime variation of NO_2 is governed by traffic [8], whose peaks happen to occur at times when the temperature is lowest. Moreover, during nighttime, the NO_2 concentration and the temperature seemed to be correlated.

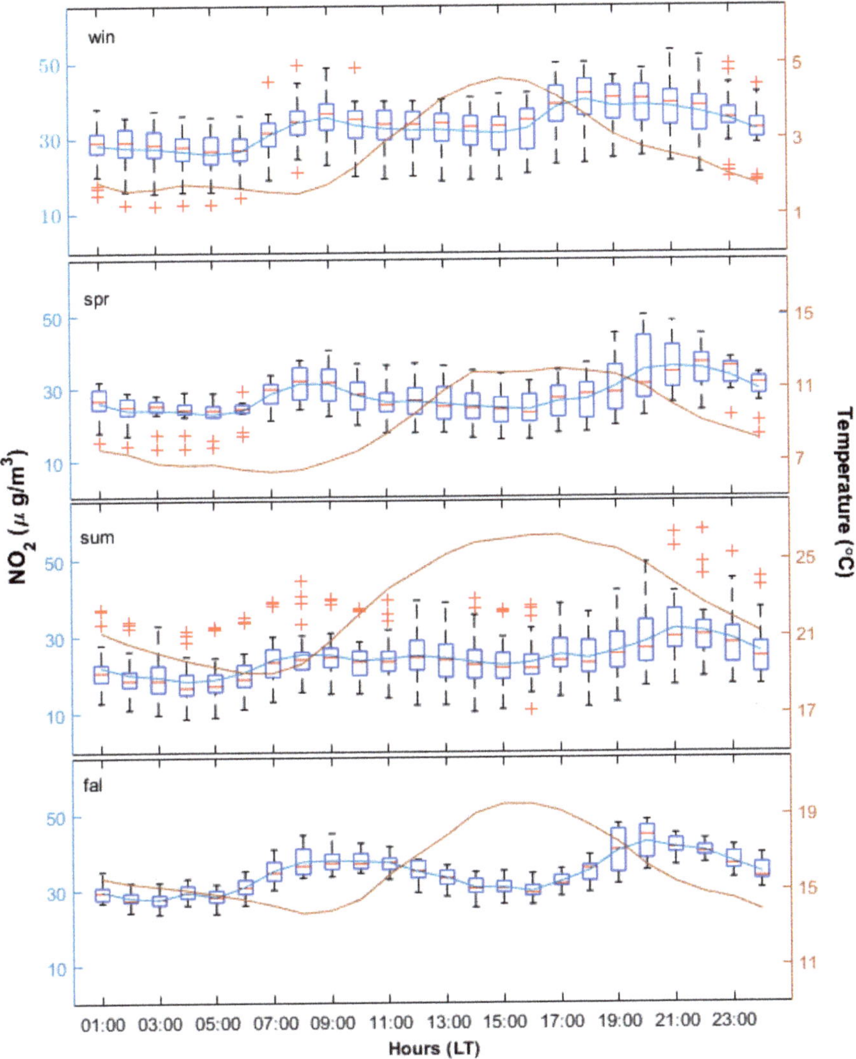

Figure 3. Diurnal variation of NO_2 concentration (boxplots, left axis) for each season (top–down: spring, summer, fall and winter) for the entire interval, 2009–2014. See text for definition of seasons. The average temperature is superimposed (orange line, right axis).

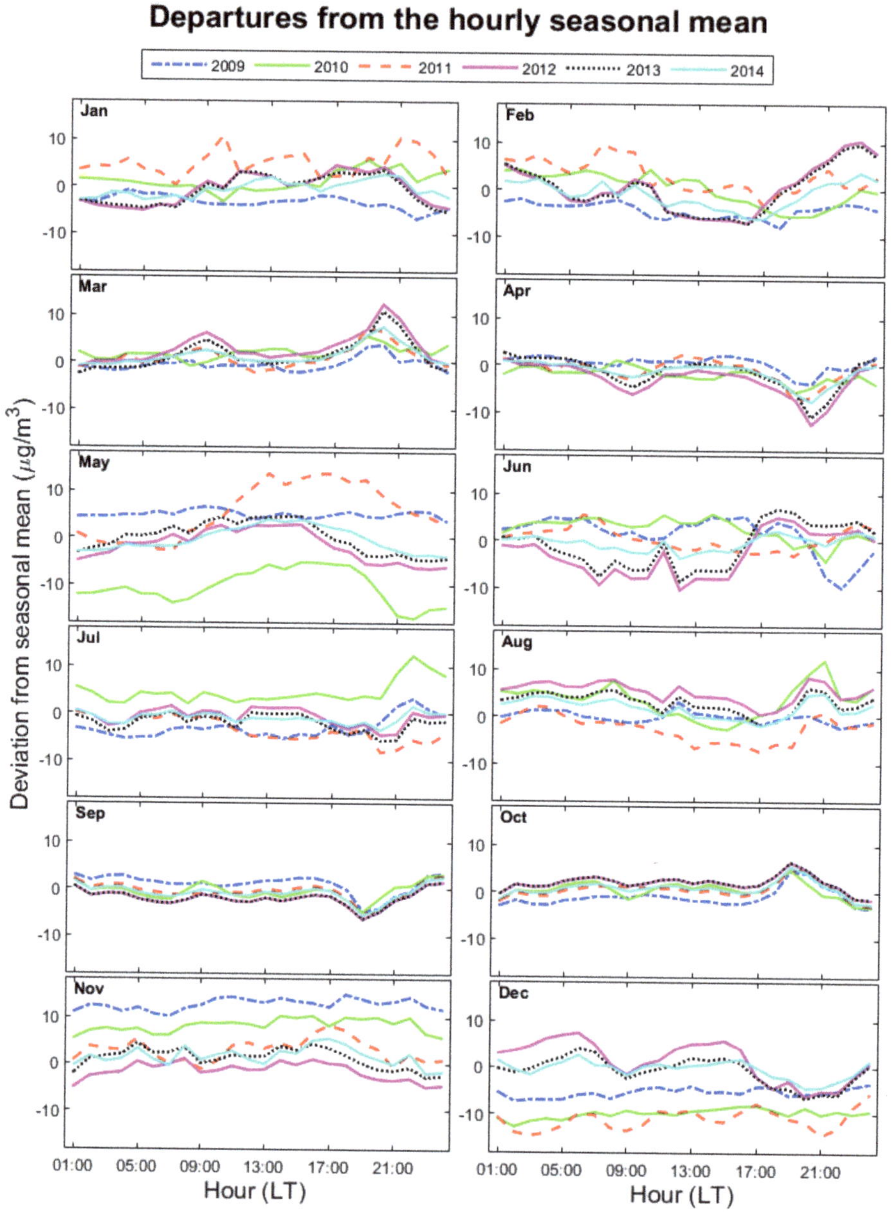

Figure 4. Departures of hourly NO_2 concentration from the hourly seasonal means for each month, January to December. Each year is shown with different colors/line styles (see legend).

3.2. Monthly Deviation of NO_2 Concentration from the Seasonal Mean

Obviously, the diurnal variation of NO_2 was not the same for each year. Figure 4 shows the difference between the hourly NO_2 concentrations at ground level and the hourly seasonal mean during each month of each year. Hourly seasonal means are averages of hourly NO_2 values measured during 2009 and 2014 over those months that define a season, i.e., November–February for winter, March–April for spring, May–August for summer and September–October for fall.

The analysis of the temporal variation of the aforementioned differences may be useful for identifying months or periods of the day when the NO_2 loading departed from the expected variability, which in turn may help in finding the cause for the outliers seen in Figure 3. Values lying in the positive/negative part mean that the NO_2 concentration were higher/lower than the average at that particular time.

A clear pattern could be observed for equinox seasons: the afternoon peak was higher during colder months (March and October) than during warmer months (April and September) for each year. This was not true for the morning peak or for solstice seasons. The month of May, which is part of the summer season, was the least regular; the NO_2 content in 2010 was much lower, while in 2011 it was much higher than the average. The unusual increase during the second part of 2010, shown in Figure 2, is confirmed by the higher values seen in Figure 4 for July 2010 and, partially, for June 2010. The explanation might relate to the construction works previously described, which resulted in significant alterations of the traffic flow during 2010–2011. In general, the major contributor to the summer mean for all years came from the NO_2 content in August, since the departures from the seasonal mean were positive for all years except 2011.

The NO_2 content in winter also depended on month and year. November seemed to be the month with the highest contribution to the winter seasonal average of NO_2 concentration for the first part of the interval (2009–2011). In December, the NO_2 diurnal variability changed with the year: the NO_2 content was lower during the first three years and higher afterwards. We assume that the explanation lies in a combination of meteorological conditions and important variations of traffic and industrial emissions during these particular years. During the analyzed period, the distribution of thermal energy and hot water in Braila municipality was mainly controlled by the heating operator SC "CET" SA. Starting with 2012, the lack of investments in modernizing the heating system affected the control of emissions: severe losses and the evolution of the methane gas tariff led to an increase of classical housing heating systems, whose impact is important especially during the cold season [39].

3.3. Correlation between NO_2 Concentration and Meteorological Parameters

The effect of meteorological parameters on the variability of NO_2 concentration for urban sites is, still, a conundrum. Table 2 shows some examples of correlations between NO_2 and meteorological factors for different locations [13–20]. The correlation between temperature and humidity, on the one side, and NO_x, on the other side, was positive, negative or insignificant. The correlation with the wind speed was more consistent, i.e., is negative for all sites, which is rather normal, since a stronger wind will disperse the NO_2 at a local urban site. This is the reason for investigating whether meteorological parameters may be linked to the variability of the NO_2 in a relatively small city, with an average level of pollution [31] and whether this relationship depends on seasons or on time of day.

Table 2. Correlations between NO_2 and meteorological parameters for various locations.

Location	Temperature	Humidity	Wind Speed	Seasons	Reference
Egypt, Cairo	Insignificant	Negative	Negative	-	[13]
India, Surat	Insignificant	-	Negative	Summer Autumn	[14]
India, Jabalpur	Negative	Negative	-	-	[17]
Saudi Arabia, Makkah	Insignificant	Negative	Negative	-	[18]
Saudi Arabia, Dhahran	Negative	Positive	Negative	Summer	[20]
China, Beijing	Negative	Positive	Negative	-	[19]
China, Shanghai	Negative	Negative	Negative	-	[19]
China, Guangzhou	Positive	Negative	Negative	-	[19]
Malaysia Kuala Lumpur	Positive	Insignificant	Negative	-	[15]
Iran, Isfahan	Negative	Negative	Negative	-	[16]

Figures 2–4 show that during about 07:00 and 21:00 LT, the NO_2 variability was strongly influenced by traffic, which is also confirmed by [8,37,40–42]. The correlation coefficients were computed separately for the full 24 h time (black bars), for daytime (8–21 LT, red bars) and for the nighttime (22–7 LT, blue bars). The separation between daytime and nighttime was done because the influence of the road traffic should be lower during nighttime, and thus meteorological parameters may play a different role in modulating the NO_2 concentration during the night. Obviously, this is not valid for radiation, which is absent during nighttime. However, one should keep in mind that the NO_2 content is largely controlled by traffic, especially during the day, and this will definitely affect the correlation with meteorological parameters (or lack thereof). However, we assumed that the traffic pressure does not change for the analysis period.

Figure 5 shows the variation of the direct bivariate correlation between NO_2 and temperature (left), and humidity (right) with time. Correlation coefficients calculated for the full day are shown by black bars, while for day (night) these are shown by red (blue) bars. Only coefficients that were significant at 90% are shown.

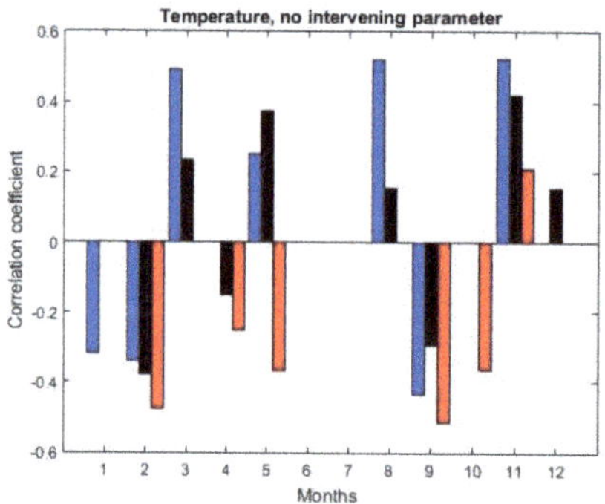

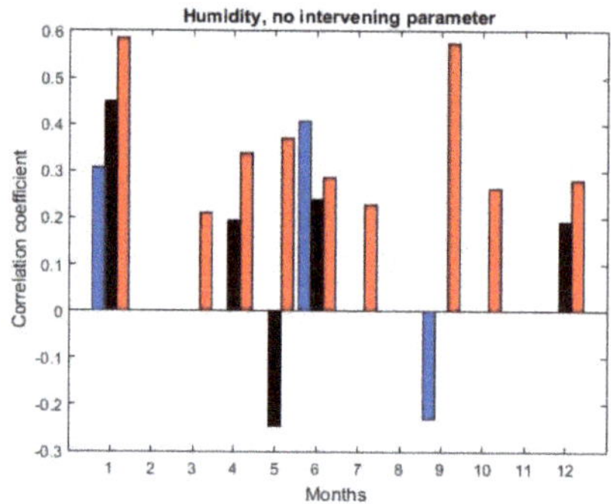

Figure 5. Full bivariate correlation of NO_2 with the temperature (**left**) and humidity (**right**). Coefficients (>90% confidence level) are shown for 24 h by black bars, while red/blue bars correspond to day/night, respectively—see text for details.

There was no clear NO_2 dependence on temperature, since correlation coefficients were positive for March, May, August and November, while for February and September these were negative. During daytime most correlations were negative; however, this was already discussed as being artificially induced by rush-hour traffic. Correlations did not change from day to night, except for May, when the correlation changed from negative for the full-day to positive for the full day and night. The full 24 h correlation was rather small or insignificant for most months and changed from positive to negative during months that had similar meteorological characteristics, e.g., March (positive), April (negative) and May (positive). Unsurprisingly, NO_2 and temperature were anticorrelated during the day, but this was already discussed as being mostly artificial. Significant correlation during the night was positive in March, May, August and November, and negative in January, February and September. All in all, temperature seemed to play no clear role in the variability of NO_2.

Negative coefficients were found for the NO_2–temperature dependence by [14,16–18], without a clear association with seasons. The NO_2 lifetime is higher during winter; thus the "night" analysis may be less relevant during winter because of the lower concentrations of the N_2O and N_2O_5 species. These species are key factors in the removal of NO_2 during nighttime and in its transformation into nitric acid [43]. The amount of NO_y species is higher during warmer seasons when bacteria and agriculture activities are intensified [44]. However, [15,19] found that an increase in temperature would be followed by an increase in NO_x.

The correlation of the NO_2 concentration with the humidity, when significant, was positive for most months. The only exception was May, during which the NO_2 variability departed significantly from the expected behavior (Figure 4). The correlation did not change from day to night except in September.

In the following the intervening effect on direct correlations is analyzed, to see whether the existing correlations were spuriously induced, especially for the link to humidity. The partial correlation was used to identify whether the effects of meteorological parameters (temperature, relative humidity and solar radiation) on NO_2 concentration were interconnected. The main and intervening variables (described in Section 2) are shown in Table 3. The results are shown only for four out of the six possible combinations, because these are the most relevant for identifying possible intervening effects. The correlation between NO_2 and temperature was inconsistent and when assessing the intervening effect of radiation on correlation with humidity one can infer that the intervening effect of humidity was similar.

Table 3. Meteorological parameters as main/intervening variables, for the partial correlation analysis.

Variable 1 (X)	Variable 2 (Z)	Intervening Variable (Y)	Plot in Figure 6
NO_2	relative humidity	temperature	a
NO_2	relative humidity	radiation	b
NO_2	radiation	temperature	c
NO_2	temperature	radiation	d

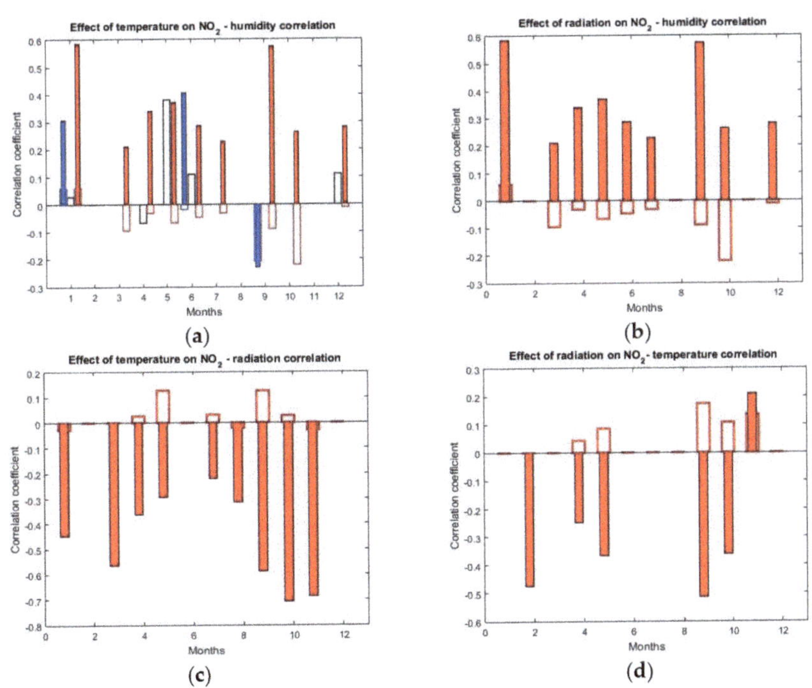

Figure 6. Partial correlation between NO_2 concentration and meteorological parameters. Full bars describe the bivariate correlation and empty bars describe the difference (*D*) between the partial and bivariate correlation. Correlation coefficients for the whole day are shown in black, while red/blue correspond to day/night, respectively—see text for details. Coefficients are shown for the >90% confidence level. Partial correlation between **NO_2 and humidity** is shown in (**a**) (*temperature* constant) and (**b**) (*radiation* constant). Partial correlation between **NO_2 and radiation** is shown in (**c**) (temperature constant) and the effect of *radiation* on the correlation with **temperature** is in (**d**).

Figure 6 shows the results of the partial correlation analysis for the combinations in Table 3. Full bars describe the direct correlation, while empty bars stand for the differences between partial and direct coefficients. According to Table 1, when these two are opposite, the correlation is partially mediated by the intervening variable. When these both have the same sign, the correlation is real.

No consistent direct correlation between NO_2 and humidity was found for the whole 24 h period (Figure 6a, full black bars are completely absent). Notably, $D_{NO(t)H}$ was rather high and positive for the whole 24 h day during May, June and December. This means that the temperature suppressed a potentially positive NO_x–humidity correlation. During daytime, NO_2 and humidity were positively correlated (red full bars) for a large part of the year, except for February, August and November. The differences between partial and direct correlation, $D_{NO(t)H}$, albeit small, were of opposing signs for most cases. This suggests that the temperature was partially responsible for the NO_2–humidity correlation.

The important role played by the humidity in the variation of NO_2 is supported by Figure 6b, where the effect of radiation on the same correlation (NO_2–humidity) is shown only for daytime. The solar radiation partially induced a positive correlation during some months, but the effect was not important, since all $D_{NO(R)H}$ were pretty small. Our results agree with [19] or [20], but contradict [13], who showed that NO_2 was negatively correlated with relative humidity. They argued that NO_2 concentrations are slightly higher at a lower relative humidity because the reactions between NO_2 and OH are less frequent, and thus the NO_2 persists more in the atmosphere. However, this was not confirmed by our results.

The NO_2 concentration was negatively correlated with solar radiation during the entire year. A higher direct radiation implies, usually, a higher air temperature; thus the intervening effect of temperature and radiation was analyzed. Figure 6c shows that the temperature did not artificially induce the anticorrelation with solar radiation (small or absent empty bars), except for May and September, when the effect was, however, small. This holds for the intervening effect of solar radiation on the anticorrelation with temperature, (Figure 6d), which was also small.

Based on observations in [45], which showed that, for a site in India, O_3 correlated negatively with both NO_x and the humidity during all seasons, we suggest that the positive correlation between NO_x and humidity may be an indirect result of the photochemical effect of solar radiation and humidity on ozone. Unfortunately, ozone measurements were not available to confirm this hypothesis. This also may partly explain the observed anticorrelation between NO_2 and solar radiation. Increased solar radiation favors the production of O_3, which, in turn, reduces the NO_2 loading in the atmosphere [45].

In general, comparisons with other studies are not straightforward, since we are not aware of similar investigations of the intervening effect of meteorological parameters. Additionally, most studies did not consider monthly changes. Moreover, correlations between the NO_2 concentration and the meteorological parameters should be different for different cities, since the anthropogenic landscape and microclimate change significantly from one urban location to another [10,31].

4. Conclusions

This paper describes the diurnal, monthly, seasonal and annual evolution of the NO_2 concentration for 2009–2014, measured in situ by an urban traffic station in southeast Romania. The role that meteorological parameters might play in modulating the NO_2 variability was investigated in an attempt to separate the anthropogenic effects (which are well-known) from the effect of the local microclimate.

As expected, the NO_2 was higher during the cold season, except for one year, 2010, when summer NO_2 levels were highest. This suggests that natural factors, such as the effect of temperature on the NO_2 lifetime, are less important than the anthropogenic ones at urban sites. Some annual variation also existed, with low values at the beginning of the interval (2009–2010), most likely caused by a severe reduction of industrial activity.The summer minima and winter maxima have both anthropogenic and natural causes and the departure from these may relate to temporary changes of the local traffic and/or construction activities. The NO_2 diurnal variability was clearly shaped by the local transport:

two diurnal peaks were observed, one around 8–9 LT and another one around 20–21 LT, and both were associated with increased road traffic, confirming previous observations at other urban sites. The afternoon peak was higher during the colder months (March and October) than during the warmer months (April and September) for each year. An irregular diurnal variation of the NO_2 concentration was seen in May and December. The most consistent season was autumn, with a relatively similar diurnal variation in all years.

Additionally, we found that over Braila, space observations of OMI followed the in situ observations during most of the selected interval (R = 0.4).

The analysis of the correlations between the NO_2 concentration and temperature, relative humidity and radiation has shown that the association with temperature is the least relevant. The correlation changed from positive to negative throughout the year without a clear pattern. Obviously, the contribution of traffic cannot be disregarded and may mask or suppress the impact of temperature variations on the NO_2 concentration. Our assumption that during the night, the situation may change due to traffic disappearance, was not confirmed. The correlations with the humidity and radiation, on the other hand, were notably consistent: the NO_2 concentration correlated with the relative humidity and was anticorrelated with radiation for almost the entire year. Moreover, most of these relationships were real and the intervening effect of the other meteorological parameters was small.

Our results showed that finding a link between meteorological parameters and NO_2 variability for an urban site is a difficult task. Attempts to predict the NO_2 behavior based on meteorological data, even combined with traffic flux data, cannot be very successful at the local or regional scale or on a short-term basis, since landscape, infrastructure, traffic, local activities, and population clearly affect the NO_2 concentration. However, this may be useful for assessing trends or long-tern variability at a large scale, since these are averaging over a large area, thus reducing local and short-term contributions.

Author Contributions: Conceptualization, M.V.; methodology, M.V. and D.-E.C.; formal analysis, M.V.; investigation, M.V. and V.C.; resources, D.-E.C., C.M.D.B., S.C.-B., V.C. and A.R.; data curation, C.M.D.B.; writing—original draft preparation, M.V.; writing—review and editing, M.V., D.-E.C., S.C.-B.; funding acquisition, M.V. All authors have read and agreed to the published version of the manuscript.

Funding: This work was supported by the EXPERT project, contract no. 14PFE/17.10.2018 with the Romanian Ministry of Research and Innovation.

Acknowledgments: We acknowledge the free use of in situ measurements of the NO_2 concentrations of the National Network for Monitoring Air Quality (RNMCA). Space-based VDC of NO_2 (OMI) were provided by the Giovanni online data system, developed and maintained by the NASA GES DIS (https://disc.gsfc.nasa.gov/datasets/OMNO2d_V003/summary).

Conflicts of Interest: The authors declare no conflict of interest.

References

1. Schaube, D.; Brunner, D.; Boersma, K.F.; Keller, J.; Folini, D.; Buchmann, B.; Berresheim, H.; Staehelin, J. SCIAMACHY tropospheric NO_2 over Switzerland: Estimates of NO_x lifetimes and impact of the complex Alpine topography on the retrieval. *Atmos. Chem. Phys.* **2007**, *7*, 5971–5987. [CrossRef]

2. Air Quality Expert Group. *Nitrogen Dioxide in the United Kingdom: Summary, prepared for The Department for Environment, Food and Rural Affairs*; Scottish Executive: Edinburgh, UK; Welsh Assembly Government: Cardiff, UK; Department of the Environment in Northern Ireland: Belfast, UK, 2004.

3. Schumann, U.; Huntrieser, H. The global lightning-induced nitrogen oxides source. *Atmos. Chem. Phys.* **2007**, *7*, 3823.

4. Stavrakou, T.; Müller, J.-F.; Boersma, K.F.; van der AR, J.; Kurokawa, J.; Ohara, T.; Zhang, Q. Key chemical NO_x sink uncertainties and how they influence top-down emissions of nitrogen oxides. *Atmos. Chem. Phys.* **2013**, *13*, 9057–9082.

5. Çelik, M.; Kadi, I. The Relation Between Meteorological Factors and Pollutants Concentrations in Karabük City. *Gazi Univ. J. Sci.* **2007**, *20*, 87–95.

6. Dunlea, E.J.; Herndon, S.C.; Nelson, D.D.; Volkamer, R.M.; San Martini, F.; Sheehy, P.M.; Zahniser, M.S.; Shorter, J.H.; Wormhoudt, J.C.; Lamb, B.K.; et al. Evaluation of nitrogen dioxide chemiluminescence monitors in a polluted urban environment. *Atmos. Chem. Phys.* **2007**, *20*, 2691–2704.

7. Kendrick, C.M.; Koonce, P.; George, L.A. Diurnal and seasonal variations of NO, NO_2 and $PM_{2.5}$ mass as a function of traffic volumes alongside an urban arterial. *Atmos. Environ.* **2015**, *122*, 133–141.

8. Dragomir, D.; Constantin, D.; Voiculescu, M.; Georgescu, L.; Merlaud, A.; Roozendael, M. Modeling results of atmospheric dispersion of NO_2 in an urban area using METI–LIS and comparison with coincident mobile DOAS measurements. *Atmos. Poll. Res.* **2015**, *6*, 503–510.

9. Cichowicz, R.; Wielgosiński, G.; Fetter, W. Dispersion of atmospheric air pollution in summer and winter season. *Environ. Monit. Assess.* **2017**, *189*, 605.

10. Degraeuwe, B.; Thunis, P.; Clappier, A.; Weiss, M.; Lefebvre, W.; Janssen, S.; Vranckx, S. Impact of passenger car NO_x emissions and NO_2 fractions on urban, NO_2 pollution e Scenario analysis for the city of Antwerp, Belgium. *Atm. Environ.* **2017**, *126*, 218–224.

11. Casquero-Vera, J.-A.; Lyamani, H.; Titos, G.; Borrás, E.; Olmo, F.; Allados-Arboledas, L. Impact of primary NO_2 emissions at different urban sites exceeding the European NO_2 standard limit. *Sci. Total Environ.* **2019**, *646*, 1117–1125. [CrossRef]

12. Parrish, D.D. Critical evaluation of US on-road vehicle emission inventories. *Atmos. Environ.* **2006**, *40*, 2288–2300.

13. Elminir, H.K. Dependence of urban air pollutants on meteorology. *Sci. Total Environ.* **2005**, *350*, 225–237.

14. Verma, S.; Desai, B. Effect of Meteorological Conditions on Air Pollution of Surat City. *J. Int. Environ. Appl. Sci.* **2008**, *3*, 358–367.

15. Dominick, D.; Latif, M.; Juahir, H.; Aris, A.; Zain, S. An assessment of influence of meteorological factors on PM_{10} and NO_2 at selected stations in Malaysia. *Sustain. Environ. Res.* **2012**, *22*, 305–315.

16. Hosseinibalam, F.; Hejazi, A. Influence of Meteorological Parameters on Air Pollution in Isfahan. *Int. Proc. Chem. Biol. Environ. Eng.* **2012**, *46*, 7–12.

17. Srivastava, R.; Sarkar, S.; Beig, G. Correlation of Various Gaseous Pollutants with Meteorological Parameter (Temperature, Relative Humidity and Rainfall). *Glob. J. Sci. Front. Res. Environ. Earth Sci.* **2014**, *14*, 57–65.

18. Habeebullah, T.M.; Munir, S.; Awad, A.H.A.; Morsy, E.A.; Seroji, A.R.; Mohammed, A.M.F. The Interaction between Air Quality and Meteorological Factors in an Arid Environment of Makkah, Saudi Arabia. *Int. J. Environ. Sci. Dev.* **2015**, *6*, 576–580. [CrossRef]

19. Zhang, H.; Wang, Y.; Hu, J.; Ying, Q.; Hu, X.-M. Relationships between meteorological parameters and criteria air pollutants in three megacities in China. *Environ. Res.* **2015**, *140*, 242–254. [CrossRef]

20. Gasmi, K.; Aljalal, A.; Al-Basheer, W.; Abdulahi, M. Analysis of NO_x, NO and NO_2 Ambient Levels as a Function of Meteorological Parameters in Dhahran, Saudi Arabia. *Air Pollut. XXV* **2017**, *1*, 77–86. [CrossRef]

21. Nascu, H.I.; Jantschi, L. *Analytical and Instrumental Chemistry, Cluj-Napoca*; Academic Press & Academic Direct: Cambridge, MA, USA, 2006; pp. 170–172.

22. Levelt, P.; Oord, G.V.D.; Dobber, M.; Malkki, A.; Visser, H.; De Vries, J.; Stammes, P.; Lundell, J.; Saari, H. The ozone monitoring instrument. *IEEE Trans. Geosci. Remote Sens.* **2006**, *44*, 1093–1101. [CrossRef]

23. Bucsela, E.; Celarier, E.; Wenig, M.; Gleason, J.; Veefkind, J.; Boersma, F.; Brinksma, E. Algorithm for NO/sub 2/ vertical column retrieval from the ozone monitoring instrument. *IEEE Trans. Geosci. Remote Sens.* **2006**, *44*, 1245–1258. [CrossRef]

24. Duncan, B.N.; Geigert, M.; Lamsal, L. A Brief Tutorial on Using the Ozone Monitoring Instrument (OMI) Nitrogen Dioxide (NO_2) Data Product for SIPS Preparation. *HAQAST Tech. Guid. Doc.* **2018**. [CrossRef]

25. Lamsal, L.N.; Martin, R.V.; Van Donkelaar, A.; Steinbächer, M.; Celarier, E.A.; Bucsela, E.; Dunlea, E.J.; Pinto, J.P. Ground-level nitrogen dioxide concentrations inferred from the satellite-borne Ozone Monitoring Instrument. *J. Geophys. Res. Space Phys.* **2008**, *113*, 1–15. [CrossRef]

26. Boersma, F.; Jacob, D.J.; Trainic, M.; Rudich, Y.; Desmedt, I.; Dirksen, R.; Eskes, H.J. Validation of urban NO_2 concentrations and their diurnal and seasonal variations observed from the SCIAMACHY and OMI sensors using in situ surface measurements in Israeli cities. *Atmos. Chem. Phys. Discuss.* **2009**, *9*, 3867–3879. [CrossRef]

27. Gruzdev, A.N.; Elokhov, A.S. Validation of Ozone Monitoring Instrument NO2 measurements using ground based NO2 measurements at Zvenigorod, Russia. *Int. J. Remote. Sens.* **2010**, *31*, 497–511. [CrossRef]

28. Constantin, D.-E.; Voiculescu, M.; Georgescu, L.P. Satellite Observations of NO_2 Trend over Romania. *Sci. World J.* **2013**, *2013*, 261634. [CrossRef]

29. Lee, H.J.; Koutrakis, P. Daily Ambient NO_2 Concentration Predictions Using Satellite Ozone Monitoring Instrument NO_2 Data and Land Use Regression. *Environ. Sci. Technol.* **2014**, *48*. [CrossRef]

30. Krotkov, N.A.; McLinden, C.A.; Li, C.; Lamsal, L.N.; Celarier, E.A.; Marchenko, S.V.; Swartz, W.H.; Bucsela, E.J.; Joiner, J.; Duncan, B.N.; et al. Aura OMI observations of regional SO_2 and NO_2 pollution changes from 2005 to 2015. *Atmos. Chem. Phys. Discuss.* **2016**, *16*, 4605–4629. [CrossRef]

31. Anand, J.S.; Monks, P.S. Estimating daily surface NO_2 concentrations from satellite data – a case study over Hong Kong using land use regression models. *Atmos. Chem. Phys.* **2017**, *17*, 8211–8230.

32. Paraschiv, S.; Constantin, D.E.; Paraschiv, S.L.; Voiculescu, M. OMI and Ground-Based In-Situ Tropospheric Nitrogen Dioxide Observations over Several Important European Cities during 2005–2014. *Int. J. Environ. Res. Public Health* **2017**, *14*, 1415. [CrossRef]

33. Krotkov, N.A. OMI/Aura NO2 Cloud-Screened Total and Tropospheric Column L3 Global Gridded 0.25° × 0.25° V3, NASA Goddard Space Flight Center, Goddard Earth Sciences Data and Information Services Center (GES DISC). Available online: https://disc.gsfc.nasa.gov/datacollection/OMNO2_CPR_003.html (accessed on 25 July 2019).

34. Usoskin, I.; Voiculescu, M.; Kovaltsov, G.; Mursula, K. Correlation between clouds at different altitudes and solar activity: Fact or Artifact? *J. Atmos. Solar-Terrestrial Phys.* **2006**, *68*, 2164–2172. [CrossRef]

35. Johnson, R.A.; Bhattacharyya, G.K. *Statistics: Principles and Methods*, 6th ed.; John Wiley & Sons, Inc.: Hoboken, NJ, USA, 2009; pp. 93–98.

36. Rosu, A.; Constantin, D.E.; Bocaneala, C.; Arseni, M.; Georgescu, L.P. Evolution of NO_2 in five major cities in Europe using remote satellite observations and in situ measurements. In *Annals of "Dunarea De Jos" University of Galati Mathematics, Physics, Theoretical Mechanics*; Ghent University: Ghent, Belgium, 2016; pp. 66–70, ISSN 20672071.

37. Constantin, D.E.; Merlaud, A.; Voiculescu, M.; Van Roozendael, M.; Arseni, M.; Rosu, A.; Georgescu, L. NO_2 and SO_2 observations in SouthEast Europe using mobile DOAS observations. *Carpath. J. Earth. Environ.* **2017**, *12*, 323–328.

38. Wang, S.; Xing, J.; Chatani, S.; Hao, J.; Klimont, Z.; Cofala, J.; Amann, M. Verification of anthropogenic emissions of China by satellite and ground observations. *Atmos. Environ.* **2011**, *45*, 6347–6358. [CrossRef]

39. S.C. TQ Consultanta & Recrutare, S.R.L. *The 2014–2020 Urban Strategy of Durable Development of Braila Town*; S.C. TQ Consultanta & Recrutare, S.R.L.: Galati, Romania, 2014; pp. 121–132. (in Romanian)

40. Rosu, A.; Rosu, B.; Arseni, M.; Constantin, D.E.; Voiculescu, M.; Georgescu, L.P.; Van Roozendael, M. Tropospheric nitrogen dioxide measurements in South-East of Romania using zenith-sky mobile DOAS observations. *New Technol. Prod. Mach. Manuf. Technol.* **2017**, *44*, 189–194.

41. Rosu, A.; Rosu, B.; Constantin, D.E.; Voiculescu, M.; Arseni, M.; Murariu, G.; Georgescu, L.P. Correlations between NO_2 distribution maps using GIS and mobile DOAS measurements in Galati city. In *Annals of "Dunarea de Jos" University of Galati Mathematics, Physics, Theoretical Mechanics*; Special Issue; Universitatea "Dunărea de Jos" din Galați: Galati, Romania, 2018; pp. 23–31. ISSN 2067207.

42. Rosu, A.; Constantin, D.; Rosu, B.; Calmuc, V.; Arseni, M.; Voiculescu, M.; Georgescu, L.P. Mobile measurements of nitrogen dioxide using two different UV-Vis spectrometers. *Tech. J. New Technol. Prod. Mach. Manuf. Technol.* **2019**, *26*, 71–76.

43. Platt, U.; Stutz, J. Differential Absorption Spectroscopy. In *Physics of Earth and Space Environments*; Springer Science and Business Media LLC: Berlin/Heidelberg, Germany, 2008; pp. 135–174.

44. Davidson, E.A.; Kingerlee, W. A global inventory of nitric oxide emissions from soils. *Nutr. Cycl. Agroecosyst.* **1997**, *48*, 37–50. [CrossRef]

45. Pancholi, P.; Kumar, A.; Bikundia, D.S.; Chourasiya, S. An observation of seasonal and diurnal behavior of O_3 – NO_x relationships and local/regional oxidant ($OX = O_3 + NO_2$) levels at a semi-arid urban site of western India. *Sustain. Environ. Res.* **2018**, *28*, 79–89. [CrossRef]

International Journal of
*Environmental Research
and Public Health*

Review

Key Points in Air Pollution Meteorology

Isidro A. Pérez *, Mª Ángeles García, Mª Luisa Sánchez, Nuria Pardo and
Beatriz Fernández-Duque

Department of Applied Physics, Faculty of Sciences, University of Valladolid, Paseo de Belén, 7,
47011 Valladolid, Spain; magperez@fa1.uva.es (M.Á.G.); mluisa.sanchez@uva.es (M.L.S.);
npardo@fa1.uva.es (N.P.); beatriz.fernandez.duque@uva.es (B.F.-D.)
* Correspondence: iaperez@fa1.uva.es

Received: 22 September 2020; Accepted: 9 November 2020; Published: 11 November 2020

Abstract: Although emissions have a direct impact on air pollution, meteorological processes may influence inmission concentration, with the only way to control air pollution being through the rates emitted. This paper presents the close relationship between air pollution and meteorology following the scales of atmospheric motion. In macroscale, this review focuses on the synoptic pattern, since certain weather types are related to pollution episodes, with the determination of these weather types being the key point of these studies. The contrasting contribution of cold fronts is also presented, whilst mathematical models are seen to increase the analysis possibilities of pollution transport. In mesoscale, land–sea and mountain–valley breezes may reinforce certain pollution episodes, and recirculation processes are sometimes favoured by orographic features. The urban heat island is also considered, since the formation of mesovortices determines the entry of pollutants into the city. At the microscale, the influence of the boundary layer height and its evolution are evaluated; in particular, the contribution of the low-level jet to pollutant transport and dispersion. Local meteorological variables have a major influence on calculations with the Gaussian plume model, whilst some eddies are features exclusive to urban environments. Finally, the impact of air pollution on meteorology is briefly commented on.

Keywords: particulate matter; atmospheric boundary layer; weather types; Gaussian plume model; low-level jet; recirculation; microscale; macroscale; mesoscale

---

1. Introduction

Air pollution is the subject of active research due to its marked impact on human life in a number of different environments, with regard to materials as well as living beings. Its impact on materials may have not only aesthetic but also economic implications. For example, the marginal costs of cleaning and repairing building façades due to air pollution have mainly been attributed to PM_{10} and SO_2 [1], while in the plant environment, a key role is played by biomonitors. *Cupressus macrocarpa* leaves were recently used to evaluate the pollution load, and high values of anthropogenic elements, such as Cu and As, were observed near an industrial area [2]. Finally, the effects on human health have also been extensively investigated due to the global scope of the air pollution problem [3]. For instance, a positive and significant association between air pollutant concentration and the incidence of breast cancer has been described by Hwang et al. [4]. Moreover, the relationship between atmospheric levels of pollutants such as particulate matter, ozone, nitrogen dioxide, and sulphur dioxide and mortality causes has been reported in Portugal [5], with the increase in health expenditure due to air pollution being the economic implication of this issue [6].

Substances released into the atmosphere become pollutants when their concentrations exceed certain levels. These values may be reached not only by the increase in emissions but also under given meteorological conditions. The main difference between the roles of emissions and meteorological conditions is that emissions may be controlled when inmission levels are exceeded [7,8], whereas meteorological conditions cannot be controlled. Consequently, air pollution control strategies should take into account meteorological variables, since these will shape any such strategies to a certain degree. Kanawade et al. [9] analysed severe air pollution episodes in New Delhi, India, which were linked to stagnant atmospheric conditions. During the study period, low-level winds were below 8 ms^{-1}, there was no precipitation, and 500 hPa wind speeds were below 13 ms^{-1}. However, meteorology may also influence air pollution transport. Dong et al. [10] analysed the seasonal impact of meteorological variations on the air quality of the Beijing–Tianjin–Hebei region, China, which is a populated metropolitan region comparable to the Pearl River Delta or the Yangtze River Delta regions, where pollution transport plays a key role. One result of this study is the successful pollution control undertaken by joint regional controls. Moreover, the noticeable impact of meteorological conditions, up to 40% on regional transport over two seasons, i.e., spring and winter, suggested that regional air pollution needed to be controlled in both seasons.

The objective of this paper is to highlight the feedback between air pollution and meteorology. Consequently, the first goal of this paper is to simplify this topic, in which a range of disciplines, such as chemistry, physics, numerical modelling, biology, medicine, or even architecture, are involved. Moreover, this paper is not only aimed at pointing out the research lines that are currently active but also at highlighting possible future studies in order to attract potential researchers to a field that stands out thanks to its social impact and constant expansion.

Although the contribution of meteorological processes to air pollution events is usually assumed, it is frequently relegated to the background. Physical processes in the atmosphere are varied and include clear-sky surface solar radiation reduction by aerosols [11] or the study of their scattering angle [12], temperature inversions associated with air pollution and health effects [13], or analysis of electricity in the convective boundary layer [14]. Moreover, meteorological processes may have contrasting effects. Meteorological conditions sometimes determine air pollution situations, since they may aggravate pollution episodes, such as the recirculation of air parcel trajectories studied in Los Angeles [15]. However, certain meteorological conditions may also improve air quality, with the inverse relationship between particulate matter concentrations and the ventilation factor (the product of wind speed at 10 m and the mixed-layer height estimate) observed in Santiago City, Chile, being one noticeable example [16].

This study differs from previous analyses in the approach used, since it takes dynamic meteorology as a basis, due to the significant weight this subject carries in atmospheric analyses. Since energy in the atmosphere flows from large eddies to molecular processes, the structure of this paper follows the scale of atmospheric motions: macroscale, mesoscale, and microscale. Macroscale considers weather phenomena with a diameter above 1000 km and with a life of several days or even weeks, such as synoptic and tropical cyclones or fronts. Mesoscale corresponds to middle-sized atmospheric structures with a diameter between 2 and 1000 km and which last from hours to a couple of days, such as mountain waves or sea breezes. Finally, microscale is linked to processes of between less than a metre and a couple of kilometres and lasting from a few seconds to several minutes. Energy exchanges at the low atmosphere together with surface conditions determine atmospheric processes at this scale, such as tornadoes, dust devils, or plumes. This classification must be considered as the result of a simplification method to separate spatial and temporal ranges. However, in agreement with atmospheric evolution, processes may sometimes appear mixed, and their influence on air pollution needs to be evaluated.

Urban environments merit special consideration since large numbers of people live, work, meet, or spend time at these sites. Urban activities are sources of pollutants, and their emissions depend on city populations and facilities. Pollutant dispersion is conditioned by urban features, and the spread of cities may vary following cultural and geographical characteristics. Consequently, the impact of cities ranges from the micro to the mesoscale, and this review considers both possibilities.

A second key point of this study is its review of the influence of air pollution on meteorology, which has scarcely been analysed to date. This impact may be noticeable at a local scale, such as the effects observed by Saha et al. [17] during the eve of the Deepawali festival at Kolkata, India, when a massive firework display triggered a marked increase in particulate matter, the consequences of which were increased surface temperature and changes in the temperature vertical profile and diurnal pattern of relative humidity. However, the effects of air pollution may also be observed at a global scale. Wang et al. [18] simulated the long-range transport of anthropogenic aerosols from Asia, which moved on the north Pacific and led to the increase in cloud top height against preindustrial conditions, resulting in an invigoration of mid-latitude cyclones.

2. Macroscale

Two groups of papers investigate the influence of meteorology at this scale on the concentrations recorded. One considers the effect of the synoptic pattern, while the second focuses on pollutant transport, which is analysed using air parcel trajectory models.

2.1. Synoptic Pattern

Atmospheric flow may be classified by discrete circulation patterns in agreement with certain modes of atmospheric circulation. Both subjective and objective procedures may be used to achieve this goal. Dayan et al. [19] highlighted the importance of these classification methods in environmental phenomena such as air pollution, desert dust intrusions, and flash floods.

Dai et al. [20] used a classification method for regional particulate pollution days in Eastern China based on a subjective approach that employs only three patterns: equalised pressure (EQP), the advancing edge of a cold front (ACF), and the inverted trough of low pressure (INT). The seasonal features of these patterns were also considered to form eight types, three EQPs (in summer, autumn, and winter), two ACFs (in autumn and winter), and three INTs (in spring, autumn, and winter). Most of the pollution days appeared in winter, with the lowest number appearing in spring. Moreover, most of the pollution days were linked to the EQP, whereas the INT was the least frequent. The main weakness of subjective procedures is that they rely on the skill and experience of the classifier, since certain atmospheric flows may display features that are compatible with different patterns.

In contrast to this relatively simple and subjective classification, objective procedures are based on mathematical expressions that could be applied not only by users who have experience in atmospheric evolution but also, in some cases, by users who are beginning their career. One example is the Lamb–Jenkinson weather-type approach, which was considered by Li et al. [21] in their study into $PM_{2.5}$ formation in North China. In this classification, the location of pressure centres was determined using gridded pressure data of a 16-point moveable region. Additionally, a small number of empirical rules was considered to classify the daily synoptic pattern as one of 26 weather types. As a result, three weather types were associated with high $PM_{2.5}$ pollution levels. Another example of an objective classification is the obliquely rotated principal component analysis in T-mode, used by Mao et al. [22] to investigate air pollution over Wuhan in wintertime from 2014 to 2018. Three of five dominant synoptic patterns were linked to heavy pollution. A third example of objective weather synoptic classification is based on the self-organising map approach. Shu et al. [23] employed this procedure to analyse summertime ozone pollution over the Yangtze River Delta in China and found six predominant synoptic weather patterns: four ozone-polluted types and two ozone-clean types.

The synoptic pattern has a noticeable impact at the microscale level, with high-pressure systems being an example. They are linked to low wind speed, quite limited expansion of the boundary layer, and a reinforced inversion layer, which thereby increases pollutant concentration [24].

2.2. Cold Fronts

Yang et al. [25] analysed the weather systems associated with nine pollution episodes in Hong Kong. Four were linked with monsoons, two with tropical cyclones, one with a saddle-type pressure field, one with a warm high pressure, and one with a cold-front passage. Cold fronts are evolving structures that have a varied impact on air pollution. Jia et al. [26] analysed the "sawtooth cycles" in Beijing, although these features have also been observed in other mid-latitude areas, such as eastern North America, that are regularly crossed by the polar front. Sawtooths are linked with double fronts, where a weak cold front precedes a second and stronger cold front. Their period is irregular, frequently around five days, although some sawtooths extend up to a week and have been confused with regular, weekly cycles. The impact of this irregular synoptic cycle on particulate matter concentration is characterised by a slow increase over a few days followed by a sharp drop in just a few hours.

Rainfall associated with cold fronts determines the decrease of particulate concentrations [27]. Cold frontal passages usually favour atmospheric pollutant removal. Hu et al. [28] observed air pollution purification during the cold frontal passage associated with a strong cold-air outbreak, since surface wind speed and atmospheric boundary layer height increased. However, there is also evidence of quite different results. The formation and evolution of air pollution episodes is influenced by wind speeds and wind shears caused by nocturnal low-level jets and cold fronts [29]. Moreover, the chemical production of sulphate, nitrate, and ammonium may increase due to the uplift of air pollutant together with humidity, which facilitates the heterogeneous and aqueous-phase oxidation of precursors [30]. Another example is presented by Kang et al. [31], who observed that particulate matter may rise to the free troposphere from the North China Plain due to a cold frontal passage. These particles were transported by strong north-westerly frontal airflow and returned to the boundary layer at the Yangtze River Delta by synoptic subsidence under the high-pressure system that dominates once the frontal zone moves downstream. They concluded that cold fronts may be seen as potential carriers of atmospheric pollutants. Similarly, Zhou et al. [32] considered the influence of large-scale cold fronts and warm conveyor belts on the transport of continental aerosols located near the surface mixed with aerosols from the upper planetary boundary layer and middle troposphere. Finally, Song et al. [33] analysed a dust storm event in Northern China in early May 2017 and concluded that the front's activity was associated with increased wind speed and dust emission.

2.3. Pollution Transport

Air pollution transport has an impact not only on people's health but also on political relationships between countries who share common borders, with particulate matter transport from Poland to the Czech Republic being one notable example when the wind direction may be established, which occurs in around 50% of cases. This is true despite the prevailing flow being from the Czech Republic to Poland, which is typical of northeast Moravia due to the orographic influence of the Moravian Gate, and though emissions from both countries contribute equally with low wind speeds, since wind direction frequently changes under such conditions. Most of the pollution was attributed to days with unclear airflow or with a significant wind change [34]. Another example is the pollution from China that affects South Korea's air quality with a seasonal evolution [35]. In this analysis, agricultural burning and coal-fired heating were noticeable in summer and winter, respectively, while dust storms from the deserts impacted in spring.

Different models have been developed to explore air parcel trajectories. The HYSPLIT (Hybrid Single-Particle Lagrangian Integrated Trajectory) model has been widely used. It was developed by the NOAA (National Oceanic and Atmospheric Administration) and, since it was first described by Draxler and Hess [36], it has been regularly updated [37]. However, other models have also gained certain relevance. The FLEXPART (FLEXible PARTicle dispersion model) may simulate long-range transport, turbulent diffusion, deposition, or linear chemistry. Since it first appeared in the mid-1990s [38], the model has also improved its performance, physical–chemical parameterisations, input and output formats, or available software [39]. Another example is the METEX (METeorological data EXplorer) model [40], whose main feature is its easy application, since only the coordinates of the site of interest are required and air parcel trajectories may be systematically calculated.

Determining air parcel trajectories is essential in terms of knowing which locations are affected by hazardous substances that have been accidentally released, such as radionuclides [41]. Air parcel trajectory models may be a necessary tool to determine the atmospheric corridors that control air flow. However, a large number of air parcel trajectories need be considered if any firm conclusions are to be reached, since handling isolated trajectories may not prove to be easy. Procedures for grouping trajectories, such as clustering analysis, are required and possible applications vary. Salmabadi et al. [42] employed the total spatial variance method to identify four main dust corridors affecting Ahvaz, Iran. Three of these, which had westerly or north-westerly orientations, were associated with the Shamal winds and were particularly active in summer. Another corridor, with a southerly orientation, was linked to a prefrontal dust-storm mechanism and was active in spring and winter. Rana et al. [43] considered the "angle distance matrix procedure" to form six trajectory groups and obtained high particulate matter concentration when the air parcels followed the route from the Middle East through the Himalayan valley to Dhaka, Bangladesh, during the dry season. Moreover, theoretical trajectory calculations must be combined with measurements and satellite imagery in order to obtain a precise knowledge of potential sources and transport times [44]. Trajectory calculations are the basis of certain useful calculations, such as the potential source contribution function and concentration-weighted trajectories [45]. In both procedures, the study region is divided into cells and the number of trajectories calculated must be high enough to obtain reliable results. The potential source contribution function considers the relationship between trajectory endpoints above a given concentration level and the total endpoints in each cell, whereas concentration-weighted trajectories calculate the average weighted concentration in each cell.

3. Mesoscale

Typical mesoscale processes are breezes induced by land–sea interactions. Orographic features condition the transport and dispersion of air pollution, and this interaction is the subject of ongoing research [46]. Moreover, air parcel recirculation and urban heat island effects also lie within this scale.

3.1. Breezes

Land–sea breezes are formed at the interface of surfaces with contrasting features. At these sites, transport of particulate matter and its composition reveal its origin. Batchvarova et al. [47] measured aerosol concentration at Ahtopol (Bulgaria) and found a lower number of coarse particles for marine air masses. Another example is the Pearl River Delta, where interactions of synoptic winds and mesoscale breezes determined weak winds and air pollution accumulation, favouring intense photochemical reactions and the formation of an ozone "pool" [48]. Similarly, in Manila, Philippines, strong sea breezes disperse fine particulate matter inland during the day in spring, although breezes are the outflow path of these aerosols during the night. However, stagnant air determines high accumulated particulate concentration in autumn when the sea breeze is weaker [49]. Dasari et al. [50] indicated that differential heating between land and sea caused strong breezes that led to diurnal dispersion and distribution of pollutants in Neom, Saudi Arabia.

Several examples illustrate the role of orographic features in pollutant transport and dispersion. Bei et al. [51] analysed the influence of mountain–valley breeze circulations in the Guanzhong basin, China, and found that this breeze determines a convergence zone in the basin where pollutants return from the mountain and concentrations increase in the evening. Alternatively, different indicators may be used to characterise the relief effect on atmospheric flow. Lai and Lin [52] analysed wind direction, the Froude number, and the pitching angle of the surface level among other variables to investigate air pollution events in Taiwan. They considered ten categories of air pollution events following the airflow and the location of the air pollution event. Moreover, the Froude number was below 0.25 in most of these events, revealing the high stability of the airflow, which tends to split around the mountains. Another noticeable effect induced by the relief was described by Quan et al. [53], who observed that a vertical vortex elevated ground pollutants at an altitude between 1.4 and 1.7 km where an elevated pollutant layer is formed and transported to Beijing by the southerly wind.

3.2. Recirculation Processes

Allwine and Whiteman [54] proposed integral quantities to describe stagnation, recirculation, and ventilation to characterise airflow at specific sites. Stagnation conditions may favour high pollutant concentrations. However, recirculation is quantified by the recirculation factor where 0 corresponds to straight-line transport without recirculation, and 1 is reached when the air parcel returns to its origin without net transport. In this latter case, concentrations may change when pollutants are affected by chemical reactions. One example is recirculation formed by sea breezes combined with upslope winds in the western Mediterranean basin during summer, and which may be considered as natural-photochemical reactors, since nitrogen oxides and precursors are transformed into oxidants, such as ozone, under strong insolation with residence times in the order of days [55]. Recirculation has also been observed in the Hong Kong area, another coastal site, where ozone pollution may be aggravated by this process [56]. Dimitriou and Kassomenos [57] analysed the impact of Saharan dust intrusions on continental Greece and concluded that certain infrequent events observed in August–September were due to dust recirculation by the Etesian winds linked to the extension of the Azores anticyclone in continental Europe. Since recirculation is a kind of eddy, orographic features, such as valleys, favour these processes. Quimbayo-Duarte et al. [58] analysed the air flow in a valley with two sections and observed recirculations when the air from the slopes converges at the valley floor and flows to the narrow section in the up-valley direction.

Another example of recirculation is the Fresno eddy [59], which is a cyclonic vortex formed during the night in the San Joaquin Valley of California. During the daytime, winds mainly enter through the San Francisco Bay area and transport precursors to the valley where ozone is formed. The air rises into the mountains at the southern end and leaves the valley. However, during the night-time, the air cannot leave the valley due to the cool wind drainage from the mountains and so returns north in a circular flow pattern, known as the Fresno eddy. Pollutants are forced to return to urban areas where more precursors are added [60]. Observations indicate that this eddy dominates ozone production in the central part of this valley and triggers many ozone exceedances if it is strong on apparently ventilated days [61].

3.3. Urban Heat Island

The features as well as the extension of certain urban structures can modify both thermal and mechanical turbulences. Absorption and emission of radiation by urban surfaces in urban environments differs from the same processes when they occur in rural surfaces. Moreover, the roughness determined by buildings is also greater. The urban heat island may enhance the existing vorticity, which produces direct thermal circulation, causing surface mesovortices. Mainhart et al. [62] observed that ozone formed outside the Saint Louis metropolitan area returned to the city due to such mesovortices.

The beneficial effect of urban heat island mitigation by urban cooling strategies on air quality is assumed. Green roofs and living walls, for example, are useful for removing air pollutants [63]. However, competing feedbacks should be considered since green and cool roofs reduce wind speed and vertical mixing during the day, which might lead to air stagnation and air pollution situations [64]. Similarly, Henao et al. [65] indicated that air pollution transport mechanisms in urban valleys could be altered by urban heat island mitigation and that air quality could worsen. In fact, Li et al. [66] presented the "urban aerosol pollution island", which is negatively correlated with the urban heat island in summer. Moreover, mitigation of the urban heat island by an increasing albedo weakened turbulent mixing, made the planetary boundary layer height decrease, and led to a stronger urban aerosol pollution island. Consequently, a balance needs to be found between urban heat island mitigation and its impact on air quality.

Droste et al. [67] introduced the "urban wind island effect" since, under certain atmospheric conditions, wind speed in the city may be higher than that corresponding to rural environments. This effect would be observed in the afternoon and would be due to the contrast between the city and the countryside in the atmospheric boundary layer growth, surface roughness, and ageostrophic wind. However, this effect has only been suggested and needs to be verified prior to gauging its influence on air pollution.

4. Microscale

At the shortest spatial scale, pollutant concentration is affected by the atmospheric boundary layer height. However, determining its depth may not prove to be easy, and the issue has been avoided in various studies where concentrations are related with meteorological variables measured at a local scale.

4.1. Atmospheric Boundary Layer

Su et al. [68] found strong interactions between the planetary boundary layer height and particulate matter concentration in winter, when the planetary boundary layer height is shallow and particulate matter concentration is high. Correlations between the planetary boundary layer height and particulate concentrations tend to be negative in general. However, the magnitude, significance, or sign of these correlations depend on the location, season, and meteorological conditions. Results indicate that this relationship is weak over clean regions but nonlinear over polluted regions.

The development of a thermal internal boundary layer at coastal sites is linked with pollution episodes. Wei et al. [69] obtained the boundary layer height with a ceilometer at Qinhuangdao, North China. Determining the thermal internal boundary layer was based on wind direction, when the wind blew from the sea, and boundary layer height. The boundary layer was thinner with a thermal internal boundary layer. If this layer was observed, particulate concentrations increased when the boundary layer height rose, since the dilution effect of the breeze is weak, and the boundary layer height was small. However, if the thermal internal boundary layer was not formed, particulate concentrations fell when the boundary layer height increased.

The atmospheric boundary layer, which develops during the day, evolves to a stable nocturnal boundary layer due to the radiative cooling of the ground and is topped by a residual layer, which is a reservoir of pollutants from daytime emissions and photochemical production. Observations indicate a wind speed maximum, known as low-level jet, near the top of the nocturnal boundary layer. This nocturnal jet may favour mixing between the two layers despite stable stratification.

The influence of low-level jets on pollutant concentrations varies. These jets may transport pollutants trapped in the residual layer and trigger episodes when mixed at the surface by convective turbulence after sunrise. One example of this process is presented by Sullivan et al. [70], who reported an ozone episode in the Baltimore–Washington DC urban corridor in June 2015, with concentrations between 70 and 100 ppb at around 1 km above the surface in the nocturnal residual layer. These concentrations were transported by the nocturnal low-level jet, and both the simulations and the measurements indicated high "next-day" ozone concentrations at sites along the southern New England region. A similar case is presented by Li et al. [71], where pollutants from the North China Plain were transported by the nocturnal low-level jet to Shenyang. Caputi et al. [72] suggested that the ozone of the residual layer is actively mixed in the nocturnal boundary layer when the low-level jet is strong. Under these conditions, ozone dry deposition causes low concentrations the following day. However, ozone is retained by the residual layer when the low-level jet is weak since the residual layer is more decoupled. In this situation, fumigation processes the following morning would determine high ozone concentrations near the ground. Wei et al. [73] also suggested an abrupt reduction in particulate matter concentration and an improvement in air quality due to enhanced vertical mixing linked to the low-level jets formed at the end of the pollution periods.

Wind speed and vertical turbulence control air pollution during the day. However, vertical turbulence is weak at night and vertical mixing is controlled by large-scale turbulence eddies whose increased number, due to low wind speed, might lead to a fall in concentration near the ground [74].

4.2. Local Meteorological Variables

Specialised devices are not usually available to most researchers. In these situations, some analyses relate pollutant concentrations with variables provided by meteorological stations.

Chen et al. [75] observed that temperature and air pressure had a marked influence on ozone concentrations in Beijing, although meteorological influence was weak in summer, since emissions of volatile organic compounds and nitrogen oxides increased in this season and led to ozone pollution episodes. The close relationship between particulate matter and meteorological variables has been the subject of several analyses. For instance, Meng et al. [76] concluded that the planetary boundary layer height and temperature difference of the inversion layer and its depth are variables that impact particulate matter concentrations. This analysis presented the ranges of these meteorological variables, where the correlation with air pollution was observed. In this line, certain thresholds of the meteorological variables should be established, since the relationship with air pollution may be the opposite. Wang et al. [77] considered 27 cities in China and studied surface solar irradiance and wind speed. Both variables displayed a similar trend between 1961 and 2011. The authors found that winds below 2.5 ms^{-1} disperse air pollutants and enhance solar surface irradiance. However, winds above 3.5 ms^{-1} enhance aerosol concentration, perhaps from dust storms, and surface solar irradiance was seen to be attenuated. Between the two thresholds, no relationship between wind speed and surface solar radiation could be established.

Local meteorology may have a marked influence when a hazardous substance is accidentally released or when an industrial facility is close to a population centre, and various mathematical models have been developed to investigate air pollutant dispersion. The Gaussian model is a classical formulation of plumes that presents a simple structure and allows multiple simulation possibilities. Moreover, it may easily be extended to more complex configurations of sources. Şahin and Ali [78] determined the emergency planning zones around two nuclear power plants using this model, and Rohmah and Nurokhim [79] simulated I^{131} dispersion around a nuclear facility. Similarly, Maués et al. [80] applied this model to evaluate the air quality of Volta Redonda, Brazil, where a large steel plant is located, and reported concentration curves around the plant showing the plume impact. Dispersion coefficients play a key role in this model. Although their calculation relies on empirical schemes, the parameterisations used may give unequal results and should be compared. Mao et al. [81] considered four dispersion schemes applied to the Prairie Grass experiments in 1956. They evaluated the performance of these schemes following the stability conditions and selected the Chinese National Standard from the results obtained for the different stability classes.

4.3. Urban Micrometeorology

Eddy formation is favoured by the rough elements and varied covers present in cities. Han et al. [82] compared the air quality between urban and rural areas in 35 major Chinese metropolitan regions. Although urban pollution is usually more marked than rural pollution in summer, both pollutions are noticeable in winter, with rural pollution being even stronger than urban pollution in certain cities. The authors found that the key factors for the longer spatial inhomogeneity in summer versus winter were emissions, urban structures, meteorology, topography, and humidity. At a smaller, daily timescale, Choi et al. [83] observed in Los Angeles that limited dispersion capacity and the relatively stable surface layer in the morning led to higher ultrafine particle concentrations than in the afternoon. In this period, wind speeds were higher and turbulence was stronger, with vertical turbulence intensity being the most effective factor controlling ultrafine particle concentrations. The effect of tall buildings on pollutant dispersion was studied by Aristodemou et al. [84], who found a concentration increase at high floors, whereas dispersion was favoured by tall buildings at low floors.

5. Influence of Air Pollution on Meteorology

Although the spread and intensity of atmospheric processes make them insensitive to external influences, the strength of air pollution might condition an area's weather or even its climate.

Wang et al. [85] studied the effect of aerosols on radiation. Surface temperature is seen to be reduced by solar radiation reduction and the temperature increase of the upper layer by solar radiation absorption or backscattering. As a result, atmospheric stratification becomes more stable. In fact, simulations with increasing aerosol optical depth have revealed a transition from unstable to stable stratification at rural sites, to weakly stable at suburban sites and to less unstable or neutral stratification at urban sites. Consequently, energy transfer from the surface is reduced, as is the boundary layer height. This impact of aerosol pollution on meteorology has been reported in numerous analyses. Nguyen et al. [86] also observed a decrease in shortwave radiation, temperature, planetary boundary layer height, and wind speed. Visibility degradation in Eastern Thailand has been studied by Aman et al. [87], who observed a marked seasonality as well as a relationship with wind direction. Moreover, this visibility degradation was linked to air pollution and meteorological conditions, such as small wet scavenging, low mixing height, and a high recirculation factor.

Evidence concerning the relationship between aerosols and precipitations is conflicting. Rosenfeld and Woodley [88] indicated that the number of cloud condensation nuclei in polluted air is over ten times higher than this number for the same volume of clean air. Consequently, under air pollution conditions, small droplets would be formed, which could not easily grow to reach precipitation size. Jirak and Cotton [89] observed this effect at elevated sites west of Denver and Colorado Springs, Colorado. However, Rosenfeld et al. [90] have explained both the decrease and the increase in rainfall by air pollution from a combination of radiative effects on surface heating and microphysical effects on invigorating the convection. Recently, Yang et al. [91] concluded that tropical cyclone precipitation was invigorated by aerosols over the Chinese mainland from 1980 to 2014. Moreover, light rain presented a decreasing trend against time, whereas the trend for heavy rains was seen to be increasing. In contrast, post-monsoon precipitation reduction in South Asia between 2004 and 2014 was explained by two factors. The first was aerosol loading, which triggered surface cooling, and the second was the temperature decrease by evaporation around irrigation hotspots [92].

6. Conclusions

Meteorological conditions have a direct influence on emission control in air pollution events. Although atmospheric processes are varied and frequently mixed, one or a few sometimes prevail and a direct relationship with high pollutant concentrations may be established.

At the synoptic scale, objective procedures provide for a systematic classification of weather types that can be linked to pollution events. However, pressure centres, which are associated with certain air mass features, move slowly and their fronts sweep the low atmosphere, with contrasting effects on air pollution. Air parcel trajectory models use meteorological observations and are a useful tool to investigate the sources, paths, and destinations of air pollution.

Air flow may be disturbed by coasts, mountains, or big cities. These features may determine horizontal or vertical vortices where pollutants may be trapped. Recirculation processes have been the subject of some studies, and their analysis continues to offer a promising line of research. Moreover, the contrast between large cities and their surrounding areas is evolving rapidly, and the corresponding changes in meteorological aspects or consequences in pollutant transport and dispersion are fields that remain open to study.

At the lowest scale, the boundary layer height is inversely correlated with pollutant concentrations. Boundary layer evolution is closely related with air pollution since a polluted residual layer may be formed at night and the low-level jet favours both pollutant transport and dispersion. The relationship between concentrations and the usual meteorological variables is common in air pollution studies. Cities are particular sites that favour turbulence and where the low atmosphere is more unstable than at rural sites. The Gaussian plume model enables pollutant concentration to be calculated and is especially useful when dealing with the release of hazardous substances. Moreover, this model may be easily adapted to greater distances and varied conditions such as trapped plumes, fumigation, or orographic features. At this scale, thresholds of meteorological variables should be investigated following air pollution levels.

Finally, the impact of air pollution on meteorology should not be excluded, with radiation absorption by aerosols being one example. The relationship between aerosols and precipitation depends on the balance between microphysical and radiative effects, and the conditions linked to this balance should be analysed.

Consequently, the examples presented in this study demonstrate that the relationship between meteorology and air pollution merits in-depth enquiry, and air pollution control strategies can no doubt benefit from this research. Although these studies may appear to be local analyses, ascertaining common features remains open to further research in order to gain insights into the feedback between meteorology and air pollution.

Author Contributions: Conceptualization, I.A.P. and M.Á.G.; methodology, I.A.P.; writing—original draft preparation, I.A.P.; writing—review and editing, N.P. and B.F.-D.; funding acquisition, M.L.S. All authors have read and agreed to the published version of the manuscript.

Funding: This review was funded by the Regional Government of Castile and Leon, project number VA027G19.

References

1.   Grøntoft, T. Estimation of Damage Cost to Building Façades per kilo Emission of Air Pollution in Norway. *Atmosphere* **2020**, *11*, 686. [CrossRef]

2.   Gorena, T.; Fadic, X.; Cereceda-Balic, F. Cupressus macrocarpa leaves for biomonitoring the environmental impact of an industrial complex: The case of Puchuncaví-Ventanas in Chile. *Chemosphere* **2020**, *260*, 127521. [CrossRef] [PubMed]

3.   Fdez-Arroyabe, P.; Robau, D.T.; De Arroyabe, P.F. Past, present and future of the climate and human health commission. *Int. J. Biometeorol.* **2017**, *61*, 115–125. [CrossRef] [PubMed]

4.   Hwang, J.; Bae, H.-J.; Choi, S.; Yi, H.; Ko, B.; Kim, N. Impact of air pollution on breast cancer incidence and mortality: A nationwide analysis in South Korea. *Sci. Rep.* **2020**, *10*, 1–7. [CrossRef] [PubMed]

5.   Brito, J.; Bernardo, A.; Zagalo, C.; Gonçalves, L.L. Quantitative analysis of air pollution and mortality in Portugal: Current trends and links following proposed biological pathways. *Sci. Total. Environ.* **2020**, *755*, 142473. [CrossRef] [PubMed]

6.   Chen, F.; Chen, Z. Cost of economic growth: Air pollution and health expenditure. *Sci. Total. Environ.* **2020**, *755*, 142543. [CrossRef] [PubMed]

7.   Cheng, N.; Cheng, B.; Li, S.; Ning, T. Effects of meteorology and emission reduction measures on air pollution in Beijing during heating seasons. *Atmos. Pollut. Res.* **2019**, *10*, 971–979. [CrossRef]

8.   Ma, T.; Duan, F.; He, K.; Qin, Y.; Tong, D.; Geng, G.; Liu, X.; Li, H.; Yang, S.; Ye, S.; et al. Air pollution characteristics and their relationship with emissions and meteorology in the Yangtze River Delta region during 2014–2016. *J. Environ. Sci.* **2019**, *83*, 8–20. [CrossRef]

9.   Kanawade, V.; Srivastava, A.; Ram, K.; Asmi, E.; Vakkari, V.; Soni, V.; Varaprasad, V.; Sarangi, C. What caused severe air pollution episode of November 2016 in New Delhi? *Atmos. Environ.* **2020**, *222*, 117125. [CrossRef]

10. Dong, Z.; Wang, S.; Xing, J.; Chang, X.; Ding, D.; Zheng, H. Regional transport in Beijing-Tianjin-Hebei region and its changes during 2014–2017: The impacts of meteorology and emission reduction. *Sci. Total. Environ.* **2020**, *737*, 139792. [CrossRef]

11. Yu, L.; Zhang, M.; Wang, L.; Lu, Y.; Li, J. Effects of aerosols and water vapour on spatial-temporal variations of the clear-sky surface solar radiation in China. *Atmos. Res.* **2020**, *248*, 105162. [CrossRef]

12. Zeng, Z.-C.; Xu, F.; Natraj, V.; Pongetti, T.J.; Shia, R.-L.; Zhang, Q.; Sander, S.P.; Yung, Y.L. Remote sensing of angular scattering effect of aerosols in a North American megacity. *Remote. Sens. Environ.* **2020**, *242*, 111760. [CrossRef]

13. Trinh, T.T.; Le, T.T.; Nguyen, T.D.H.; Minh, T.B. Temperature inversion and air pollution relationship, and its effects on human health in Hanoi City, Vietnam. *Environ. Geochem. Health* **2019**, *41*, 929–937. [CrossRef] [PubMed]

14. Anisimov, S.; Galichenko, S.; Prokhorchuk, A.; Aphinogenov, K. Mid-latitude convective boundary-layer electricity: A study by large-eddy simulation. *Atmos. Res.* **2020**, *244*, 105035. [CrossRef]

15. Jury, M.R. Meteorology of air pollution in Los Angeles. *Atmos. Pollut. Res.* **2020**, *11*, 1226–1237. [CrossRef]

16. Muñoz, R.C.; Corral, M.J. Surface Indices of Wind, Stability, and Turbulence at a Highly Polluted Urban Site in Santiago, Chile, and their Relationship with Nocturnal Particulate Matter Concentrations. *Aerosol Air Qual. Res.* **2017**, *17*, 2780–2790. [CrossRef]

17. Saha, U.; Talukdar, S.; Jana, S.; Maitra, A. Effects of air pollution on meteorological parameters during Deepawali festival over an Indian urban metropolis. *Atmos. Environ.* **2014**, *98*, 530–539. [CrossRef]

18. Wang, Y.; Wang, M.; Zhang, R.; Ghan, S.J.; Lin, Y.; Hu, J.; Pan, B.; Levy, M.; Jiang, J.H.; Molina, M.J. Assessing the effects of anthropogenic aerosols on Pacific storm track using a multiscale global climate model. *Proc. Natl. Acad. Sci. USA* **2014**, *111*, 6894–6899. [CrossRef]

19. Dayan, U.; Tubi, A.; Levy, I. On the importance of synoptic classification methods with respect to environmental phenomena. *Int. J. Clim.* **2011**, *32*, 681–694. [CrossRef]

20. Dai, Z.; Liu, D.; Yu, K.; Cao, L.; Jiang, Y. Meteorological Variables and Synoptic Patterns Associated with Air Pollutions in Eastern China during 2013–2018. *Int. J. Environ. Res. Public Health* **2020**, *17*, 2528. [CrossRef]

21. Li, M.; Wang, L.; Liu, J.; Gao, W.; Song, T.; Sun, Y.; Li, L.; Li, X.; Wang, Y.; Liu, L.; et al. Exploring the regional pollution characteristics and meteorological formation mechanism of PM2.5 in North China during 2013–2017. *Environ. Int.* **2020**, *134*, 105283. [CrossRef] [PubMed]

22. Mao, F.; Zang, L.; Wang, Z.; Pan, Z.; Zhu, B.; Gong, W. Dominant synoptic patterns during wintertime and their impacts on aerosol pollution in Central China. *Atmos. Res.* **2020**, *232*, 104701. [CrossRef]

23. Shu, L.; Wang, T.; Han, H.; Xie, M.; Chen, P.; Li, M.; Wu, H. Summertime ozone pollution in the Yangtze River Delta of eastern China during 2013–2017: Synoptic impacts and source apportionment. *Environ. Pollut.* **2020**, *257*, 113631. [CrossRef] [PubMed]

24. Sun, X.; Wang, K.; Li, B.; Zong, Z.; Shi, X.; Ma, L.; Fu, D.; Thapa, S.; Qi, H.; Tian, C. Exploring the cause of PM2.5 pollution episodes in a cold metropolis in China. *J. Clean. Prod.* **2020**, *256*, 120275. [CrossRef]

25. Yang, Y.; Yim, S.H.L.; Haywood, J.; Osborne, M.; Chan, C.S.; Zeng, Z.; Cheng, J.C. Characteristics of Heavy Particulate Matter Pollution Events Over Hong Kong and Their Relationships With Vertical Wind Profiles Using High-Time-Resolution Doppler Lidar Measurements. *J. Geophys. Res. Atmos.* **2019**, *124*, 9609–9623. [CrossRef]

26. Jia, Y.; Rahn, K.A.; He, K.; Wen, T.; Wang, Y. A novel technique for quantifying the regional component of urban aerosol solely from its sawtooth cycles. *J. Geophys. Res. Space Phys.* **2008**, *113*. [CrossRef]

27. Andreão, W.L.; Trindade, B.T.; Nascimento, A.P.; Reis, N.C.; Andrade, M.F.; Albuquerque, T.T.D.A. Influence of Meteorology on Fine Particles Concentration in Vitória Metropolitan Region During Wintertime. *Rev. Bras. Meteorol.* **2019**, *34*, 459–470. [CrossRef]

28. Hu, Y.; Wang, S.; Ning, G.; Zhang, Y.; Wang, J.; Shang, Z. A quantitative assessment of the air pollution purification effect of a super strong cold-air outbreak in January 2016 in China. *Air Qual. Atmos. Health* **2018**, *11*, 907–923. [CrossRef]

29. Li, X.; Ma, Y.; Wang, Y.; Wei, W.; Zhang, Y.; Liu, N.; Hong, Y. Vertical Distribution of Particulate Matter and its Relationship with Planetary Boundary Layer Structure in Shenyang, Northeast China. *Aerosol Air Qual. Res.* **2019**, *19*, 2464–2476. [CrossRef]

30. Wang, T.; Huang, X.; Wang, Z.; Liu, Y.; Zhou, D.; Ding, K.; Wang, H.; Qi, X.; Ding, A. Secondary aerosol formation and its linkage with synoptic conditions during winter haze pollution over eastern China. *Sci. Total. Environ.* **2020**, *730*, 138888. [CrossRef]

31. Kang, H.; Zhu, B.; Gao, J.; He, Y.; Wang, H.; Su, J.; Pan, C.; Zhu, T.; Yu, B. Potential impacts of cold frontal passage on air quality over the Yangtze River Delta, China. *Atmos. Chem. Phys. Discuss.* **2019**, *19*, 3673–3685. [CrossRef]

32. Zhou, D.; Ding, K.; Huang, X.; Liu, L.; Liu, Q.; Xu, Z.; Jiang, F.; Fu, C.; Ding, A. Transport, mixing and feedback of dust, biomass burning and anthropogenic pollutants in eastern Asia: A case study. *Atmos. Chem. Phys. Discuss.* **2018**, *18*, 16345–16361. [CrossRef]

33. Song, P.; Fei, J.; Li, C.; Huang, X. Simulation of an Asian Dust Storm Event in May 2017. *Atmosphere* **2019**, *10*, 135. [CrossRef]

34. Černikovský, L.; Krejčí, B.; Blažek, Z.; Volná, V. Transboundary Air-Pollution Transport in the Czech-Polish Border Region between the Cities of Ostrava and Katowice. *Cent. Eur. J. Public Health* **2016**, *24*, S45–S50. [CrossRef] [PubMed]

35. Kim, M.J. The effects of transboundary air pollution from China on ambient air quality in South Korea. *Heliyon* **2019**, *5*. [CrossRef]

36. Draxler, R.R.; Hess, G.D. An overview of the HYSPLIT_4 modelling system for trajectories, dispersion and deposition. *Aust. Meteorol. Mag.* **1998**, *47*, 295–308.

37. Stein, A.F.; Draxler, R.R.; Rolph, G.D.; Stunder, B.J.B.; Cohen, M.D.; Ngan, F. NOAA's HYSPLIT Atmospheric Transport and Dispersion Modeling System. *Bull. Am. Meteorol. Soc.* **2015**, *96*, 2059–2077. [CrossRef]

38. Stohl, A.; Hittenberger, M.; Wotawa, G. Validation of the lagrangian particle dispersion model FLEXPART against large-scale tracer experiment data. *Atmos. Environ.* **1998**, *32*, 4245–4264. [CrossRef]

39. Pisso, I.; Sollum, E.; Grythe, H.; Kristiansen, N.I.; Cassiani, M.; Eckhardt, S.; Arnold, D.; Morton, D.; Thompson, R.L.; Zwaaftink, C.D.G.; et al. The Lagrangian particle dispersion model FLEXPART version 10.4. *Geosci. Model. Dev.* **2019**, *12*, 4955–4997. [CrossRef]

40. Zeng, J.; Matsunaga, T.; Mukai, H. METEX—A flexible tool for air trajectory calculation. *Environ. Model. Softw.* **2010**, *25*, 607–608. [CrossRef]

41. Hernández-Ceballos, M.; Sangiorgi, M.; García-Puerta, B.; Montero, M.; Trueba, C. Dispersion and ground deposition of radioactive material according to airflow patterns for enhancing the preparedness to N/R emergencies. *J. Environ. Radioact.* **2020**, *216*, 106178. [CrossRef] [PubMed]

42. Salmabadi, H.; Khalidy, R.; Saeedi, M. Transport routes and potential source regions of the Middle Eastern dust over Ahvaz during 2005–2017. *Atmos. Res.* **2020**, *241*, 104947. [CrossRef]

43. Rana, M.; Mahmud, M.; Khan, M.H.; Sivertsen, B.; Sulaiman, N. Investigating Incursion of Transboundary Pollution into the Atmosphere of Dhaka, Bangladesh. *Adv. Meteorol.* **2016**, *2016*, 1–11. [CrossRef]

44. Rogers, H.M.; Ditto, J.C.; Gentner, D.R. Evidence for impacts on surface-level air quality in the northeastern US from long-distance transport of smoke from North American fires during the Long Island Sound Tropospheric Ozone Study (LISTOS) 2018. *Atmos. Chem. Phys. Discuss.* **2020**, *20*, 671–682. [CrossRef]

45. Bao, Z.; Chen, L.; Li, K.; Han, L.; Wu, X.; Gao, X.; Azzi, M.; Cen, K. Meteorological and chemical impacts on PM2.5 during a haze episode in a heavily polluted basin city of eastern China. *Environ. Pollut.* **2019**, *250*, 520–529. [CrossRef]

46. De Wekker, S.F.; Kossmann, M.; Knievel, J.C.; Giovannini, L.; Gutmann, E.D.; Zardi, D. Meteorological Applications Benefiting from an Improved Understanding of Atmospheric Exchange Processes over Mountains. *Atmosphere* **2018**, *9*, 371. [CrossRef]

47. Batchvarova, E.; Calidonna, C.; Kolarova, M.; Ammoscato, I.; Barantiev, D.; Hristova, E.; Kirova, H.; Neykova, R.; Savov, P.; Kolev, N.; et al. Meteorology and air pollution experiment at the Black Sea coastal site Ahtopol—2017. *AIP Conf. Proc.* **2019**, *2075*, 120001. [CrossRef]

48. Zeren, Y.; Guo, H.; Lyu, X.; Jiang, F.; Wang, Y.; Liu, X.; Zeng, L.; Li, M.; Li, L. An Ozone "Pool" in South China: Investigations on Atmospheric Dynamics and Photochemical Processes Over the Pearl River Estuary. *J. Geophys. Res. Atmos.* **2019**, *124*, 12340–12355. [CrossRef]

49. Bagtasa, G.; Yuan, C.-S. Influence of local meteorology on the chemical characteristics of fine particulates in Metropolitan Manila in the Philippines. *Atmos. Pollut. Res.* **2020**, *11*, 1359–1369. [CrossRef]

50. Dasari, H.P.; Desamsetti, S.; Langodan, S.; Karumuri, R.K.; Singh, S.; Hoteit, I. Atmospheric conditions and air quality assessment over NEOM, kingdom of Saudi Arabia. *Atmos. Environ.* **2020**, *230*. [CrossRef]

51. Bei, N.; Zhao, L.; Xiao, B.; Meng, N.; Feng, T. Impacts of local circulations on the wintertime air pollution in the Guanzhong Basin, China. *Sci. Total. Environ.* **2017**, *592*, 373–390. [CrossRef] [PubMed]

52. Lai, H.-C.; Lin, M.-C. Characteristics of the upstream flow patterns during PM2.5 pollution events over a complex island topography. *Atmos. Environ.* **2020**, *227*, 117418. [CrossRef]

53. Quan, J.; Dou, Y.; Zhao, X.; Liu, Q.; Sun, Z.; Pan, Y.; Jia, X.; Cheng, Z.; Ma, P.; Su, J.; et al. Regional atmospheric pollutant transport mechanisms over the North China Plain driven by topography and planetary boundary layer processes. *Atmos. Environ.* **2020**, *221*, 117098. [CrossRef]

54. Allwine, K.; Whiteman, C. Single-station integral measures of atmospheric stagnation, recirculation and ventilation. *Atmos. Environ.* **1994**, *28*, 713–721. [CrossRef]

55.	Millán, M.M.; Sanz, M.J.; Salvador, R.; Mantilla, E. Atmospheric dynamics and ozone cycles related to nitrogen deposition in the western Mediterranean. *Environ. Pollut.* **2002**, *118*, 167–186. [CrossRef]

56.	Wang, H.; Lyu, X.; Guo, H.; Wang, Y.; Zou, S.; Ling, Z.; Wang, X.; Jiang, F.; Zeren, Y.; Pan, W.; et al. Ozone pollution around a coastal region of South China Sea: Interaction between marine and continental air. *Atmos. Chem. Phys. Discuss.* **2018**, *18*, 4277–4295. [CrossRef]

57.	Dimitriou, K.; Kassomenos, P. Estimation of North African dust contribution on PM10 episodes at four continental Greek cities. *Ecol. Indic.* **2019**, *106*, 105530. [CrossRef]

58.	Quimbayo-Duarte, J.; Staquet, C.; Chemel, C.; Arduini, G. Impact of Along-Valley Orographic Variations on the Dispersion of Passive Tracers in a Stable Atmosphere. *Atmosphere* **2019**, *10*, 225. [CrossRef]

59.	Lin, Y.-L.; Jao, I.-C. A Numerical Study of Flow Circulations in the Central Valley of California and Formation Mechanisms of the Fresno Eddy. *Mon. Weather Rev.* **1995**, *123*, 3227–3239. [CrossRef]

60.	2004 Extreme Ozone Attainment Demonstration Plan. Available online: http://www.valleyair.org/Air_Quality_Plans/AQ_Final_Adopted_Ozone2004.htm (accessed on 21 September 2020).

61.	Beaver, S.; Palazoglu, A. Influence of synoptic and mesoscale meteorology on ozone pollution potential for San Joaquin Valley of California. *Atmos. Environ.* **2009**, *43*, 1779–1788. [CrossRef]

62.	Mainhart, M.; Pasken, R.; Chiao, S.; Roark, M. Surface mesovortices in relation to the urban heat island effect over the Saint Louis metropolitan area. *Urban. Clim.* **2020**, *31*, 100580. [CrossRef]

63.	Viecco, M.; Vera, S.; Jorquera, H.; Bustamante, W.; Gironás, J.; Dobbs, C.; Leiva, E. Potential of Particle Matter Dry Deposition on Green Roofs and Living Walls Vegetation for Mitigating Urban Atmospheric Pollution in Semiarid Climates. *Sustainability* **2018**, *10*, 2431. [CrossRef]

64.	Sharma, A.; Conry, P.; Fernando, H.J.S.; Hamlet, A.F.; Hellmann, J.J.; Chen, F. Green and cool roofs to mitigate urban heat island effects in the Chicago metropolitan area: Evaluation with a regional climate model. *Environ. Res. Lett.* **2016**, *11*, 064004. [CrossRef]

65.	Henao, J.J.; Rendón, A.M.; Salazar, J.F. Trade-off between urban heat island mitigation and air quality in urban valleys. *Urban. Clim.* **2020**, *31*, 100542. [CrossRef]

66.	Li, H.; Sodoudi, S.; Liu, J.; Tao, W. Temporal variation of urban aerosol pollution island and its relationship with urban heat island. *Atmos. Res.* **2020**, *241*, 104957. [CrossRef]

67.	Droste, A.M.; Steeneveld, G.J.; Holtslag, A.A.M. Introducing the urban wind island effect. *Environ. Res. Lett.* **2018**, *13*, 094007. [CrossRef]

68.	Su, T.; Li, Z.; Kahn, R.A. Relationships between the planetary boundary layer height and surface pollutants derived from lidar observations over China: Regional pattern and influencing factors. *Atmos. Chem. Phys. Discuss.* **2018**, *18*, 15921–15935. [CrossRef]

69.	Wei, J.; Tang, G.; Zhu, X.; Wang, L.; Liu, Z.; Cheng, M.; Münkel, C.; Li, X.; Wang, Y.S. Thermal internal boundary layer and its effects on air pollutants during summer in a coastal city in North China. *J. Environ. Sci.* **2018**, *70*, 37–44. [CrossRef]

70.	Sullivan, J.T.; Rabenhorst, S.D.; Dreessen, J.; McGee, T.J.; Delgado, R.; Twigg, L.; Sumnicht, G. Lidar observations revealing transport of O3 in the presence of a nocturnal low-level jet: Regional implications for "next-day" pollution. *Atmos. Environ.* **2017**, *158*, 160–171. [CrossRef]

71.	Li, X.; Hu, X.-M.; Ma, Y.; Wang, Y.; Li, L.; Zhao, Z. Impact of planetary boundary layer structure on the formation and evolution of air-pollution episodes in Shenyang, Northeast China. *Atmos. Environ.* **2019**, *214*. [CrossRef]

72.	Caputi, D.J.; Faloona, I.C.; Trousdell, J.; Smoot, J.; Falk, N.; Conley, S. Residual layer ozone, mixing, and the nocturnal jet in California's San Joaquin Valley. *Atmos. Chem. Phys. Discuss.* **2019**, *19*, 4721–4740. [CrossRef]

73.	Wei, W.; Zhang, H.S.; Cai, X.; Song, Y.; Bian, Y.; Xiao, K.; Zhang, H. Influence of Intermittent Turbulence on Air Pollution and Its Dispersion in Winter 2016/2017 over Beijing, China. *J. Meteorol. Res.* **2020**, *34*, 176–188. [CrossRef]

74.	Li, X.; Gao, C.Y.; Gao, Z.; Zhang, X. Atmospheric boundary layer turbulence structure for severe foggy haze episodes in north China in December 2016. *Environ. Pollut.* **2020**, *264*, 114726. [CrossRef] [PubMed]

75.	Chen, Z.; Zhuang, Y.; Xie, X.; Chen, D.; Cheng, N.; Yang, L.; Li, R. Understanding long-term variations of meteorological influences on ground ozone concentrations in Beijing During 2006–2016. *Environ. Pollut.* **2019**, *245*, 29–37. [CrossRef] [PubMed]

76.	Meng, C.; Cheng, T.; Gu, X.; Shi, S.; Wang, W.; Wu, Y.; Bao, F. Contribution of meteorological factors to particulate pollution during winters in Beijing. *Sci. Total. Environ.* **2019**, *656*, 977–985. [CrossRef] [PubMed]

77. Wang, Y.W.; Yang, Y.H.; Zhou, X.Y.; Zhao, N.; Zhang, J.H. Air pollution is pushing wind speed into a regulator of surface solar irradiance in China. *Environ. Res. Lett.* **2014**, *9*, 054004. [CrossRef]
78. Şahin, S.; Ali, M. Emergency Planning Zones Estimation for Karachi-2 and Karachi-3 Nuclear Power Plants using Gaussian Puff Model. *Sci. Technol. Nucl. Install.* **2016**, *2016*, 1–8. [CrossRef]
79. Rohmah, R.N. Nurokhim Simulation of I-131 Dispersion around KNS (Kawasan Nuklir Serpong) using Gaussian Plume Model. *Int. Conf. Qual. Res.* **2017**, 414–419. [CrossRef]
80. Maués, C.S.; Guimarães, C.S.; Martinez-Amariz, A. Study for the dispersion of particulate matter emissions from a steel industry using Gaussian Plume equation through computational modeling. *J. Phys. Conf. Ser.* **2019**, *1386*. [CrossRef]
81. Mao, S.; Lang, J.; Chen, T.; Cheng, S.; Cui, J.; Shen, Z.; Hu, F. Comparison of the impacts of empirical power-law dispersion schemes on simulations of pollutant dispersion during different atmospheric conditions. *Atmos. Environ.* **2020**, *224*, 117317. [CrossRef]
82. Han, W.; Li, Z.; Guo, J.; Su, T.; Chen, T.; Wei, J.; Cribb, M. The Urban–Rural Heterogeneity of Air Pollution in 35 Metropolitan Regions across China. *Remote. Sens.* **2020**, *12*, 2320. [CrossRef]
83. Choi, W.; Ranasinghe, D.; Bunavage, K.; DeShazo, J.; Wu, L.; Seguel, R.; Winer, A.M.; Paulson, S.E. The effects of the built environment, traffic patterns, and micrometeorology on street level ultrafine particle concentrations at a block scale: Results from multiple urban sites. *Sci. Total. Environ.* **2016**, *553*, 474–485. [CrossRef] [PubMed]
84. Aristodemou, E.; Mottet, L.; Constantinou, A.; Pain, C.C. Turbulent Flows and Pollution Dispersion around Tall Buildings Using Adaptive Large Eddy Simulation (LES). *Buildings* **2020**, *10*, 127. [CrossRef]
85. Wang, X.; He, X.; Miao, S.-G.; Dou, Y. Numerical simulation of the influence of aerosol radiation effect on urban boundary layer. *Sci. China Earth Sci.* **2018**, *61*, 1844–1858. [CrossRef]
86. Nguyen, G.T.H.; Shimadera, H.; Sekiguchi, A.; Matsuo, T.; Kondo, A. Investigation of aerosol direct effects on meteorology and air quality in East Asia by using an online coupled modeling system. *Atmos. Environ.* **2019**, *207*, 182–196. [CrossRef]
87. Aman, N.; Manomaiphiboon, K.; Pengchai, P.; Suwanathada, P.; Srichawana, J.; Assareh, N. Long-Term Observed Visibility in Eastern Thailand: Temporal Variation, Association with Air Pollutants and Meteorological Factors, and Trends. *Atmosphere* **2019**, *10*, 122. [CrossRef]
88. Rosenfeld, D.; Woodley, W. Pollution and clouds. *Phys. World* **2001**, *14*, 33–38. [CrossRef]
89. Jirak, I.L.; Cotton, W.R. Effect of Air Pollution on Precipitation along the Front Range of the Rocky Mountains. *J. Appl. Meteorol. Clim.* **2006**, *45*, 236–245. [CrossRef]
90. Rosenfeld, D.; Lohmann, U.; Raga, G.B.; O'Dowd, C.D.; Kulmala, M.; Fuzzi, S.; Reissell, A.; Andreae, M.O. Flood or Drought: How Do Aerosols Affect Precipitation? *Science* **2008**, *321*, 1309–1313. [CrossRef]
91. Yang, X.; Zhou, L.; Zhao, C.; Yang, J. Impact of aerosols on tropical cyclone-induced precipitation over the mainland of China. *Clim. Chang.* **2018**, *148*, 173–185. [CrossRef]
92. Chen, W.-T.; Huang, K.-T.; Lo, M.-H.; Ho, L. Post-Monsoon Season Precipitation Reduction over South Asia: Impacts of Anthropogenic Aerosols and Irrigation. *Atmosphere* **2018**, *9*, 311. [CrossRef]

Publisher's Note: MDPI stays neutral with regard to jurisdictional claims in published maps and institutional affiliations.

MDPI

St. Alban-Anlage 66

4052 Basel

Switzerland

Tel. +41 61 683 77 34

Fax +41 61 302 89 18

www.mdpi.com

International Journal of Environmental Research and Public Health Editorial Office

E-mail: ijerph@mdpi.com

www.mdpi.com/journal/ijerph